Calculus for Business and Economics

Calculus for Business and Economics: An Example-Based Introduction is designed for first-year university students specializing in business and economics. This book is written in a clear, easy-to-read style and covers all the essential calculus-related topics that students are likely to encounter in their studies. With real-world business and economics applications seamlessly integrated around the core calculus concepts, students will find the book to be of real practical value throughout their time in university and beyond.

Features
- Three hundred easy-to-follow examples throughout, carefully crafted to illustrate the concepts and ideas discussed.
- Numerous exercises for practice, with solutions available online to help you learn at your own pace.
- Each chapter concludes with a section showcasing the real-world business and economics applications of the discussed mathematical concepts.

Calculus for Business and Economics
An Example-Based Introduction

Jon Pierre Fortney and Linda Smail

CRC Press
Taylor & Francis Group
Boca Raton London New York

CRC Press is an imprint of the
Taylor & Francis Group, an **informa** business

A CHAPMAN & HALL BOOK

Designed cover image: ShutterStock Images

First edition published 2025
by CRC Press
2385 NW Executive Center Drive, Suite 320, Boca Raton FL 33431

and by CRC Press
4 Park Square, Milton Park, Abingdon, Oxon, OX14 4RN

CRC Press is an imprint of Taylor & Francis Group, LLC

ISBN: 978-1-032-76430-6 (hbk)
ISBN: 978-1-032-76850-2 (pbk)
ISBN: 978-1-003-48023-5 (ebk)

DOI: 10.1201/9781003480235

Typeset in Latin Modern font
by KnowledgeWorks Global Ltd.

Publisher's note: This book has been prepared from camera-ready copy provided by the authors.

*To all my family and friends who have
supported and been there for me over the years:
my parents Dan and Marlene,
my sister Heather,
and Rene, Roger, Rajesh, and Yohann.*

*To my loving mother, Khedidja, for her
endless support and love,
my father, Cherif, whose memory
continues to inspire me,
and my brother, Amine, for his
endless support.*

Contents

Preface

Calculus for Business and Economics: An Example-Based Introduction is a calculus book exclusively intended for first-year university students majoring in business and economics. As such, this book is unique in several important ways. Unlike most calculus books currently used for business calculus classes, this book exclusively focuses on business and economics applications. This focus allows for an exceptionally clean, succinct, and targeted presentation of the mathematical material, resulting in an easy-to-read, concise, and in-depth presentation. Each chapter concludes with a section illustrating the practical business and economics applications of the covered mathematical concepts.

Topics covered include cost/revenue/profit functions, marginal functions, price-demand and supply functions, interest rate formulas, nominal and effective interest rates, diminishing marginal returns, consumer and producer surplus, a wide range of business and economics-related examples that look at marketing and sales situations, natural resource depletion, depreciation, loan repayments, and probability functions, along with the elasticity of demand, the Cobb-Douglas production function, and maximizing/minimizing functions subject to constraints.

Many introductory-level textbooks have ballooned to the point that any actual expectation that students read or work from the textbook has become unreasonable. Addressing this issue has been the primary driving force behind this book. It is far shorter than other business calculus books on the market. But despite this, it covers somewhat more material, featuring topics often omitted from traditional business calculus textbooks such as integration by substitution, parts, and partial fractions, the extreme value theorem, implicit differentiation, multivariable functions, partial derivatives, and Lagrange multipliers. This is partially accomplished by a very careful, straightforward, and concise use of language. Sentences and paragraphs are kept short and easy to understand without loss of mathematical rigor. Explanations are kept simple, short, and clear.

Another key aspect of this book is that only the essential aspects of each mathematical concept are covered. This is a math book, but one with a very specific audience, and this audience needs the mathematical tools necessary to analyze and understand business and economics concepts. Thus, we avoid any mathematical aspects that are not essential to the goal at hand. For example, we cover limits within the context of the definition of the derivative, and then only to the extent necessary for the topic at hand, making this a particularly user-friendly textbook that is carefully crafted to meet students' needs and aid them in understanding the critical mathematical concepts for their field.

Finally, the prose is interspersed with over three hundred easy-to-follow examples. Examples are boxed for easy reference and are carefully crafted to illustrate the concepts and ideas discussed. The examples are an integral aspect of the presentation. They serve the dual purposes of helping students understand the material as well as serving as models for problem-solving.

Jon Pierre Fortney
Linda Smail
2024

Author Bios

Jon Pierre Fortney graduated from the University of Pennsylvania in 1996 with a B.A. in Mathematics and Actuarial Science and a B.S.E. in Chemical Engineering. Prior to returning to graduate school, he worked as both an actuarial analyst and an environmental engineer. He graduated from Arizona State University in 2008 with a Ph.D. in Mathematics, specializing in Geometric Mechanics. He has almost 20 years of university-level teaching experience, 10 of which were in the Middle East. Dr. Fortney is currently a faculty member at the University of Maryland Baltimore County (UMBC).

Dr. Fortney is devoted to teaching and pedagogy. He has published two undergraduate-level mathematics textbooks: *A Visual Introduction to Differential Forms and Calculus on Manifolds* and *Discrete Mathematics for Computer Science: An Example-Based Introduction*. He has taught business calculus for almost 15 years, which serves as the impetus for this current book.

Linda Smail graduated from the University of Batna, Algeria, in 1998 with a B.S. in Mathematics. She then earned an M.S. in Analysis and Applied Mathematics from Marne-La-Vallée University, France, and completed her Ph.D. in Applied Mathematics at the same institution in 2004, specializing in Bayesian Networks. With over 20 years of university-level teaching experience, 15 of which were in the Middle East, Dr. Smail has demonstrated a strong commitment to education and research. In 2017, she furthered her education by earning a B.S. in Mathematics and Economics from the London School of Economics.

Dr. Smail has developed and taught numerous courses in business mathematics, economics, and related fields, significantly enhancing the curriculum to better meet the needs of students. She has published numerous research papers in reputable journals and presented her work at international conferences. Her research interests include applied mathematics, Bayesian networks, and their applications in economics and business. She is currently a faculty member at Zayed University, where she continues to contribute to the academic community through teaching, research, and mentorship.

CHAPTER 1

Basic Algebra

To be successful in calculus, it is essential that you have a solid understanding of a number of basics from algebra. Our intention in this chapter is to quickly review some of the basic concepts and ideas that will be extremely helpful in the rest of this book.

We will cover the fundamentals of number sets and number properties, exponents, fractions, order of operations, polynomials, functions, and graphs of functions. When reviewing functions, we will emphasize, in particular, the functions and the fundamental tools necessary for business and economics applications.

1.1 NUMBER SETS AND PROPERTIES

Since algebra is, at a fundamental level, based on the properties of numbers, we will begin this chapter by looking at the number systems that play a role in calculus. For this, it is helpful to have a basic understanding of the language and notations of set theory.

A **set** is a **well-defined** collection of objects. The objects in a set are usually called the **elements** of the set. The phrase *well-defined* just means that it is clear if an object is in the set or not. We use capital letters to denote sets and small letters for the elements. Sets are usually indicated by curly brackets { }. A set is defined either by a list of its elements or by a rule which defines the elements of the set. Thus the set $A = \{1, 2, 3\}$ is the collection of numbers 1, 2, and 3. The numbers 1, 2, and 3 are called the elements of the set. The set $B = \{x \mid x^3 - 1 = 0\}$ is the set of one element 1. B is written using what is sometimes called "set-builder" notation. The vertical bar | is read out loud as "such that." Thus B is the set of elements x such that x is a solution to the equation $x^3 - 1 = 0$, which is, in this case, only the element 1. A set without any elements is denoted by $\emptyset$ and called the **empty set** or the **null set**. In order to indicate if an element belongs to a set, we use the symbol $\in$; for example, we have $3 \in A$. If an element does not belong to a set, we use the symbol $\notin$; for example, $5 \notin A$.

DOI: 10.1201/9781003480235-1

Probably the most important sets in mathematics are the following sets of numbers:

$$\mathbb{N} = \text{the set of } \textbf{natural numbers}$$
$$= \{1, 2, 3, 4, \ldots\},$$
$$\mathbb{Z} = \text{the set of } \textbf{integer numbers}$$
$$= \{\ldots, -4, -3, -2, -1, 0, 1, 2, 3, 4, \ldots\},$$
$$\mathbb{Q} = \text{the set of } \textbf{rational numbers}$$
$$= \left\{ x \mid x = \frac{m}{n} \text{ where } m, n \in \mathbb{Z} \text{ and } n \neq 0 \right\},$$
$$\mathbb{R} = \text{the set of } \textbf{real numbers}.$$

Sometimes the natural numbers are also called the **whole numbers** or **counting numbers**. In some books the set of natural numbers includes zero as well. Also, the integer numbers are often simply called the **integers**, the rational numbers are often simply called the **rationals**, and the real numbers are often simply called the **reals**.

Notice the set of rational numbers is written using the set-builder notation. Thus the set of rational numbers is the set of x such that x is a fraction of two integers. Of course, the integer in the denominator cannot be zero. In essence, the rational numbers are simply the set of all fractions. It is difficult to give a precise definition of the real numbers, but you should be familiar with them as the real number line. That is a good way to think about the real numbers.

If A and B are sets, the set B is said to be a **subset** of set A if every element of B is also an element of A. Thus we can think of the subset B as being "contained" inside A. We would write this as $B \subset A$ or $B \subseteq A$. The symbol $\subseteq$ allows the possibility that the two sets are actually equal. By inspection, it is fairly easy to see that the set of natural numbers is contained inside the set of integer numbers, which is itself contained inside the set of rational numbers, which is itself contained inside the set of real numbers. We would write this as

$$\mathbb{N} \subset \mathbb{Z} \subset \mathbb{Q} \subset \mathbb{R}.$$

Intervals are often used in mathematics and are an important concept in calculus. Intervals are subsets of the set of real numbers. Let x and y be two real numbers such that $x < y$. The intervals are noted as sets as follows:

$$[x, y] = \{a \in \mathbb{R} \mid x \leq a \leq y\} \subset \mathbb{R}$$
$$(x, y) = \{a \in \mathbb{R} \mid x < a < y\} \subset \mathbb{R}$$
$$[x, y) = \{a \in \mathbb{R} \mid x \leq a < y\} \subset \mathbb{R}$$
$$(x, y] = \{a \in \mathbb{R} \mid x < a \leq y\} \subset \mathbb{R}$$

Thus, the interval $[x, y]$ consists of both x and y and all the real numbers between x and y. Similarly, the interval (x, y) consists of only the real numbers between x and y, and so on. The closed brackets [or] are used when the appropriate endpoint is included in the interval, while the open brackets (or) are used when it is not included.

Numbers have certain properties. Whether or not you know the actual names of the properties, you certainly know the properties themselves. Consider $7 + 3$. We all know that this is exactly the same as $3 + 7$. The order in which we add two numbers does not matter. This is called the **commutativity of addition**. Similarly, 3×7 is exactly the same as 7×3. This is called the **commutativity of multiplication**.

Example 1.1

Commutative Property of Addition:

- $21 + 16 = 16 + 21$,
- $x + y = y + x$.

Commutativity Property of Multiplication:

- $6 \times 7 = 7 \times 6$,
- $x \times y = y \times x$. When we are using variables, we usually do not write down the $\times$ sign, so this could also be written as $xy = yx$.

Suppose we wanted to add $5 + 7 + 9$. The order in which we add these numbers does not matter. We could first do the addition $5 + 7$ to get 12, and then add 9 to get 21. Using parenthesis to indicate which addition is done first, we would write this as $(5 + 7) + 9$. Or we could first add $7 + 9$ to get 16 and then add 5 to get 21. Using parenthesis to indicate which addition is done first, we would write this as $5 + (7 + 9)$. This is called the **associativity of addition**. Similarly, $(5 \times 7) \times 9$ is exactly the same as $5 \times (7 \times 9)$, which is called the **associativity of multiplication**.

Example 1.2

Associative Property of Addition:

- $(13 + 5) + 10 = 13 + (5 + 10)$,
- $(x + y) + z = x + (y + z)$.

Associative Property of Multiplication:

- $(4 \times 8) \times 10 = 4 \times (8 \times 10)$,
- $(x \times y) \times z = x \times (y \times z)$. Again, when using variables we usually do not write down the $\times$ sign, so this could also be written as $(xy)z = x(yz)$.

Next, notice that $7 + 0 = 7$. In fact, when we add zero to any number at all we get that number back. Thus, zero is called the **additive identity**. Similarly, $9 \times 1 = 9$. When we multiply any number at all by one we get that number back. Thus, one is called the **multiplicative identity**.

Example 1.3

Identity Element of Addition:

- $16 + 0 = 16$,

- $x + 0 = x$.

Identity Element of Multiplication:

- $21 \times 1 = 21$. Using parenthesis to represent multiplication we can write $21(1) = 21$.

- $x \times 1 = x$, which could also be written as $x(1) = x$.

Finally, notice that $5 + (-5) = 0$. If we add the negative of any number to that number, we get zero, the additive identity. We call the negative of any number the **additive inverse** of that number. Similarly, notice that $6 \times \left(\frac{1}{6}\right) = 1$. If we multiply the **reciprocal** of any number with that number we get one, the multiplicative identity. The reciprocal of a number is simply one divided by that number. We call the reciprocal of any number the **multiplicative inverse** of that number.

Example 1.4

Inverse of Addition:

- $13 + (-13) = 0$,

- $-5 + \left(-(-5)\right) = 0$,

- $x + (-x) = 0$.

Inverse of Multiplication:

- $21 \times \frac{1}{21} = 1$, which could be written $21 \left(\frac{1}{21}\right) = 1$.

- $x \times \frac{1}{x} = 1$, which could be written $x \left(\frac{1}{x}\right) = 1$.

There is one more property that is extremely important in mathematics, the **distributive property**. The distributive property says we can "distribute multiplication over addition." This is illustrated in the next example.

Example 1.5

Distributive Property:

- $7 \times (5 + 3) = (7 \times 5) + (7 \times 3)$,
- $x(y + z) = xy + xz$.

Notice, because of the commutative property of multiplication, we also have

- $(x + y)z = xz + yz$.

Being aware of these properties is extremely important since they are the foundations of arithmetic and algebra. Many of the algebraic manipulations that you are probably familiar with depend on these properties.

1.2 EXPONENTS, FRACTIONS, AND ORDER OF OPERATIONS

We begin this section by looking at **exponents**. Exponents tell us to multiply a number by itself a certain number of times. For example, in the expression 5^3 the 3 is the exponent, and it tells us to multiply five by itself three times, so $5^3 = 5 \times 5 \times 5$. Sometimes the symbol $\cdot$ is used to represent $\times$, so we could write $5^3 = 5 \cdot 5 \cdot 5$. Given 5^3, we would say that "5 is **raised to the** 3rd power" or that "5 is raised to the power 3."

Example 1.6

Examples of exponents:

- $8^4 = 8 \cdot 8 \cdot 8 \cdot 8 = 4096$,
- $(-2)^5 = (-2) \cdot (-2) \cdot (-2) \cdot (-2) \cdot (-2) = -32$,
- $x^9 = x \cdot x \cdot x \cdot x \cdot x \cdot x \cdot x \cdot x \cdot x$,
- $x^1 = x$.

The identity $x^1 = x$ is sometimes called the identity rule of exponents. In reality, it is simply a notational convention, when the exponent is one, it is usually simply not written. Numbers can also have negative exponents. Any number raised to the negative first power is simply the **reciprocal** of that number. The reciprocal of a number is simply one divided by that number. Thus the reciprocal of 5 is $\frac{1}{5}$. So we have the negative exponent rule

$$x^{-1} = \frac{1}{x}$$

for any number $x \neq 0$. (Recall, division by zero is undefined.) In general, we have

$$x^{-n} = \left(\frac{1}{x}\right)^n.$$

Example 1.7

Examples of negative exponents:

- $4^{-3} = \left(\frac{1}{4}\right)^3 = \frac{1}{4} \cdot \frac{1}{4} \cdot \frac{1}{4} = \frac{1}{4 \cdot 4 \cdot 4} = \frac{1}{4^3} = \frac{1}{64}$,
- $x^{-4} = \left(\frac{1}{x}\right)^4 = \frac{1}{x} \cdot \frac{1}{x} \cdot \frac{1}{x} \cdot \frac{1}{x} = \frac{1}{x \cdot x \cdot x \cdot x} = \frac{1}{x^4}$,
- $x^{-n} = \frac{1}{x^n}$.

Exponentiation follows certain rules. First, we show the **product rule of exponentiation**. Suppose we wanted to find $7^3 \cdot 7^5$, then we would have

$$7^3 \cdot 7^5 = \overbrace{7 \cdot 7 \cdot 7}^{3 \text{ times}} \cdot \overbrace{7 \cdot 7 \cdot 7 \cdot 7 \cdot 7}^{5 \text{ times}} = 7^8 = 7^{3+5}.$$

This is true in general, giving us the following product rule:

$$x^n \cdot x^m = x^{n+m}$$

for any $x \in \mathbb{R}$ and $m, n \in \mathbb{N}$.

Next, we look at the **quotient rule of exponentiation**. Suppose we wanted to find $\frac{8^6}{8^4}$. Recalling that $1 \cdot x = x$, we have

$$\frac{8^6}{8^4} = \frac{\overbrace{8 \cdot 8 \cdot 8 \cdot 8 \cdot 8 \cdot 8}^{6 \text{ times}}}{\underbrace{1 \cdot 8 \cdot 8 \cdot 8 \cdot 8}_{4 \text{ times}}} = \frac{\not{8} \cdot \not{8} \cdot \not{8} \cdot \not{8} \cdot 8 \cdot 8}{1 \cdot \not{8} \cdot \not{8} \cdot \not{8} \cdot \not{8}} = \frac{\overbrace{8 \cdot 8}^{6 - 4 \text{ times}}}{1} = 8^2.$$

This is true in general giving the following quotient rule:

$$\frac{x^n}{x^m} = x^{n-m}.$$

Again, here we have $x \in \mathbb{R}$ and $m, n \in \mathbb{N}$.

Now we look at the **power rule of exponentiation**. Suppose we wanted to find $(8^2)^3$. It should be obvious that

$$(8^2)^3 = (8^2) \cdot (8^2) \cdot (8^2) = 8 \cdot 8 \cdot 8 \cdot 8 \cdot 8 \cdot 8 = 8^6 = 8^{2(3)}.$$

In general, we have the following power rule:

$$(x^m)^n = x^{mn}.$$

The **negative rule of exponentiation** has already been stated; it is the last item in the example of negative exponents. The negative rule of exponents is given by

$$x^{-n} = \frac{1}{x^n}.$$

Finally, we look at the **exponent of zero rule**. Suppose we wanted to find x^0 for some $x \neq 0$. Notice for any x, where $x \neq 0$, that $\frac{x^n}{x^n} = x^{n-n} = x^0$ by the quotient rule and also that $\frac{x^n}{x^n} = \frac{\not{x^n}}{\not{x^n}} = \frac{1}{1} = 1$. So, combining this we get that $x^0 = 1$. This holds whenever $x \neq 0$ since we are not allowed to divide by zero. But mathematicians have also decided to define 0^0 as 1. This means that for any value of x, we have the following exponent of zero rule:

$$x^0 = 1.$$

Example 1.8

Using the rules of exponentiation to simplify expressions:

- $(3x^3)(4x^5) = 12x^8$
- $\frac{15x^{10}}{5x^3} = 3x^7$
- $(6x^5)^2 = 36x^{10}$
- $2x^{-3} = \frac{2}{x^3}$

There are a couple of other properties that may be useful to notice. Consider the following example:

$$(2 \cdot 3)^4 = (2 \cdot 3)(2 \cdot 3)(2 \cdot 3)(2 \cdot 3) = (2 \cdot 2 \cdot 2 \cdot 2)(3 \cdot 3 \cdot 3 \cdot 3) = 2^4 \cdot 3^4.$$

Of course, we could have computed $(2 \cdot 3)^4 = 6^4$, but we wanted to make clear the following property, which is called the **power of a product rule of exponentiation**:

$$(x \cdot y)^n = x^n \cdot y^n.$$

Something similar happens when we are dealing with fractions, the following happens:

$$\left(\frac{2}{3}\right)^4 = \left(\frac{2}{3}\right)\left(\frac{2}{3}\right)\left(\frac{2}{3}\right)\left(\frac{2}{3}\right) = \frac{2^4}{3^4}.$$

In general, we have the following **power of a quotient rule of exponentiation**:

$$\left(\frac{x}{y}\right)^n = \frac{x^n}{y^n}.$$

Finally, we want to make clear some standard exponential notation that is used with roots. The square root of x is usually written as $\sqrt{x}$ or sometimes as $\sqrt[2]{x}$. We have that

$$\sqrt{x} = \sqrt[2]{x} = x^{\frac{1}{2}}.$$

For a general n^{th}-root, we have

$$\sqrt[n]{x} = x^{\frac{1}{n}}.$$

How would we write an expression like $\sqrt[n]{x^m}$, where under the nth-root, we have x^m. This becomes the **fractional exponent rule**,

$$\sqrt[n]{x^m} = x^{\frac{m}{n}}.$$

Example 1.9

Using the rules of exponentiation to simplify expressions:

- $(4x)^3 = 4^3 x^3$
- $\left(\frac{7}{2x}\right)^5 = \frac{7^5}{2^5 x^5}$
- $\sqrt[3]{x^2} = x^{\frac{2}{3}}$
- $\frac{1}{\sqrt[4]{x^3}} = x^{-\frac{3}{4}}$

We summarize the exponential rules below. These rules are frequently used in simplifying algebraic expressions and should be committed to memory.

Exponential Rules: Suppose m and n are real numbers, then

- $x^m \cdot x^n = x^{m+n}$
- $(xy)^m = x^m y^m$
- $x^0 = 1$
- $\frac{x^m}{x^n} = x^{m-n}$
- $x^{-m} = \frac{1}{x^m}$
- $(x^m)^n = x^{mn}$
- $\left(\frac{x}{y}\right)^m = \frac{x^m}{y^m}$
- $x^{\frac{m}{n}} = \sqrt[n]{x^m}$

Now we turn our attention to **fractions**. Fractions are sometimes described as "parts of a whole" and are usually represented using two numbers, one above the other, separated by a line which represents division. The number on the top is called the **numerator** and the number on the bottom is called the **denominator**. Some examples of fractions are $\frac{4}{5}$, $\frac{1}{2}$, 7/10, and 3/8. Notice the two different styles of writing fractions that are typically used.

In order to add or subtract two fractions, the denominator of both fractions must be the same,

$$\frac{a}{b} + \frac{c}{b} = \frac{a+c}{b}$$

and

$$\frac{a}{b} - \frac{c}{b} = \frac{a-c}{b}.$$

If two fractions have the same denominator, they are said to have a **common denominator**. If we want to add or subtract fractions that do not have a common denominator, we need to first obtain a common denominator. This can be done by multiplying both the numerator and denominator of the first fraction by the denominator of the second fraction, and then multiplying both the numerator and denominator of the second fraction by the denominator of the first fraction,

$$\frac{a}{b} + \frac{c}{d} = \frac{ad}{bd} + \frac{cb}{db} = \frac{ad+bc}{bd}$$

and

$$\frac{a}{b} - \frac{c}{d} = \frac{ad}{bd} - \frac{cb}{db} = \frac{ad-bc}{bd}.$$

Example 1.10

Addition and subtraction of fractions.

- $\frac{3}{4} + \frac{2}{4} = \frac{3+2}{4} = \frac{5}{4}$
- $\frac{3}{4} - \frac{2}{4} = \frac{3-2}{4} = \frac{1}{4}$
- $\frac{4}{5} + \frac{8}{3} = \frac{4\cdot3}{5\cdot3} + \frac{8\cdot5}{3\cdot5} = \frac{12+40}{15} = \frac{52}{15}$
- $\frac{4}{5} - \frac{8}{3} = \frac{4\cdot3}{5\cdot3} - \frac{8\cdot5}{3\cdot5} = \frac{12-40}{15} = \frac{-28}{15}$

When we multiply two fractions, we simply multiply the two numerators and multiply the two denominators,

$$\frac{a}{b} \times \frac{c}{d} = \frac{ac}{bd}.$$

When we divide one fraction by another fraction, we multiply the first fraction by the reciprocal of the fraction we are dividing by,

$$\frac{a/b}{c/d} = \frac{a}{b} \times \frac{d}{c} = \frac{ad}{bc}.$$

Example 1.11

Multiplying and dividing fractions.

- $\frac{2}{3} \times \frac{5}{7} = \frac{2\cdot5}{3\cdot7} = \frac{10}{21}$
- $\frac{2/3}{5/7} = \frac{2}{3} \times \frac{7}{5} = \frac{2\cdot7}{3\cdot5} = \frac{14}{15}$
- $\frac{-3}{5} \times \frac{2}{7} = \frac{-3\cdot2}{5\cdot7} = \frac{-6}{35}$
- $\frac{-3/5}{2/7} = \frac{-3}{5} \times \frac{7}{2} = \frac{-3\cdot7}{5\cdot2} = \frac{-21}{10}$

Fractions in which the numerator is greater than or equal to the denominator are called **improper fractions** and fractions in which the numerator is smaller than the denominator are called **proper fractions**. From the last two examples, $\frac{5}{4}$, $\frac{52}{15}$, $\frac{-28}{15}$, and $\frac{-21}{10}$ are all improper fractions while $\frac{1}{4}$, $\frac{10}{21}$, $\frac{14}{15}$, and $\frac{-6}{35}$ are all proper fractions. Improper fractions can also be written as **mixed fractions**. Mixed fractions have a "whole number" part and a "proper fraction" part. When the numerator is divided by the denominator, the quotient becomes the "whole number" part, the remainder becomes the numerator of the "proper fraction" part, and the original denominator becomes the denominator for the "proper fraction" part.

Example 1.12

Writing improper fractions as mixed fractions.

- $\frac{5}{4} = 1\frac{1}{4}$
- $\frac{52}{15} = 3\frac{7}{15}$
- $\frac{-28}{15} = -\frac{28}{15} = -1\frac{13}{15}$
- $\frac{-21}{10} = -\frac{21}{10} = -2\frac{1}{10}$

We now turn our attention to the **order of operations**. The order of operations is really nothing more than a *convention*. That means that mathematicians simply decided on what order the operations should be done in and that was what was used. Having an agreed-upon order of operations makes communicating much easier. Notice there are three different ways to write parenthesis, ( ), [ ], or { }.

P – Parenthesis	( ), [ ], { }	Inside parenthesis
E – Exponents	$a^x, \sqrt{x}$	All exponents
M or **D** – Multiplication or Division	$\times, \div$	Left → Right
A or **S** – Addition or Subtraction	$+, -$	Left → Right

All the multiplications and divisions are done at the same time, going left to right, as are the additions and subtractions. Clearly, "Left → Right" means from Left to Right.

The phrase I used to memorize the order of operations was "Please Excuse My Dear Aunt Sally." Sometimes the acronym PEMDAS is also used. Sometimes the word bracket is used instead of the word parenthesis, and the acronym BEDMAS is used.

Example 1.13

Find $7 + \frac{12}{4} - 2 \cdot 3$.

Solution:

$$7 + \frac{12}{4} - 2 \cdot 3 \qquad \text{(No parenthesis or exponents.)}$$
$$\implies 7 + 3 - \mathbf{2 \cdot 3} \qquad \text{(Do M or D first, L} \rightarrow \text{R, so do D first.)}$$
$$\implies 7 + \mathbf{3} - \mathbf{6} \qquad \text{(Then we do the M second.)}$$
$$\implies 10 - 6 \qquad \text{(Then do A or S, L} \rightarrow \text{R, so do A first.)}$$
$$\implies 4 \qquad \text{(Then do S second.)}$$

The symbol $\implies$ is a mathematical symbol that means "implies" or "it follows." When you see it you should understand that what is written before the $\implies$ is the same thing as what is written after the $\implies$ even if it looks different. Using the $\implies$ as it is used here is one important way to keep your work neat and to help you avoid making mistakes.

Example 1.14

Find $2 - \frac{(12-8)^2}{2} + 6 \cdot 2$.

Solution:

$$2 - \frac{(12 - 8)^2}{2} + 6 \cdot 2 \qquad \text{(Do what is in parenthesis first.)}$$
$$\implies 2 - \frac{(4)^2}{2} + 6 \cdot 2 \qquad \text{(Then do exponents.)}$$
$$\implies 2 - \frac{16}{2} + 6 \cdot 2 \qquad \text{(Do M or D first, L} \rightarrow \text{R, so do D first.)}$$
$$\implies 2 - 8 + 6 \cdot 2 \qquad \text{(Then we do the M second)}$$
$$\implies 2 - 8 + 12 \qquad \text{(Then do A or S, L} \rightarrow \text{R, so do S first.)}$$
$$\implies -6 + 12 \qquad \text{(Then do A second.)}$$
$$\implies 6$$

The actual reason that mathematicians decided on this order of operations is that it allows one to write down polynomials in a concise expanded form without using lots of parenthesis. Consider the following polynomial:

$$5x^4 + 7x^3 - 2x^2 + 3x.$$

The order of operations is implicit in this expression. We all know that the exponents are calculated first, then all the multiplications are done, and finally, the additions and subtractions. If mathematicians had chosen a different order of operations, then polynomials likes this would have to be written down in a more complicated way.

1.3 POLYNOMIAL BASICS

We begin this section by defining what a polynomial is, and then spend some time explaining what the definition actually means. A **polynomial** is an expression consisting of variables and coefficients which contains only the operations of addition, subtraction, multiplication, division, and non-negative integer exponentiation. There is a lot in this definition to unpack.

You have probably learned the difference between a sentence and a phrase. A sentence is a group of words that express a complete thought and a phrase is a group of words that does not express a complete thought. In mathematics an **expression** is a lot like a phrase that stands for a number. Some examples of expressions are

$$3x^2 - 2x + 7, \qquad -5xy^3 + 4x^3y^2, \qquad -2(6xz - 4xy - 2yz).$$

An **equation** is a sentence that relates two things with a relation symbol, with the most common being $=$. Just like sentences use phrases, equations use expressions. Some examples of equations are

$$3x^2 - 2x + 7 = 0, \qquad -5xy^3 + 4x^3y^2 = 100, \qquad -2(6xz - 4xy - 2yz) = 8xy.$$

In summary, we have the following:

Expression	Equation
Phrase	Sentence
Has no relation symbol	Has a relation symbol $(=, <, \leq, >, \geq)$
One *simplifies* an expression	One *solves* an equation
Example: $2(x+3)$	Example: $2(x+3) = 16$

Consider the expression $3x^2 - 2x + 7$. It consists of the variable x, the coefficients 3, 2, and 7, additions and subtractions, multiplications (the coefficient 3 is multiplied by x^2 and the coefficient 2 is multiplied by x), and non-negative integer exponentiation (x is raised to the power 2). Thus it meets our definition for a polynomial expression.

Example 1.15

Some examples of expressions:

- $3x^2 - 2x + 7$
  Notice, we could also have written this expression as $3x^2 + (-2)x + 7$.

Thinking of "subtraction" as "adding a negative" is so common that people often do it without thinking. This polynomial expression has three **terms**, $3x^3$, $-2x$, and 7. The terms of a polynomial are separated by additions.

- $-25xy + 42x^2y + 15xy^2 - 36x^2y^2$
  This polynomial has four terms, $-25xy$, $42x^2y$, $15xy^2$, and $-36x^2y^2$.

Terms in polynomial expressions that have the same variables which are raised to the same powers are called **like terms**. So, $3x^2y$ and $-5x^2y$ are like terms because they have the same variables (x and y) which are raised to the same powers. (Both x's are raised to the second power and both y's are raised to the first power.)

If we are given a polynomial expression that has like terms, we can simplify the expression by combining the like terms. For example, the expression $3x^2y - 5x^2y$ is combined by adding or subtracting the coefficients. Thus $3x^2y - 5x^2y$ simplifies to $-2x^2y$. The below series of examples illustrate how to simplify expressions.

Example 1.16

Simplify the expression $2x - 3xy + 4yx + 5y - 7x + 2y + x$. Since multiplication is commutative, we have $4yx = 4xy$.

Solution:
$$2x - 3xy + 4yx + 5y - 7x + 2y + x$$
$$\implies 2x - 7x + x + 5y + 2y - 3xy + 4xy$$
$$\implies -4x + 7y + xy$$

Example 1.17

Simplify the expression $4a^2b - 2ab^2 - 3a^2b^2 + 7ab^2 - 4a^2b^2 - a^2b$.

Solution:
$$4a^2b - 2ab^2 - 3a^2b^2 + 7ab^2 - 4a^2b^2 - a^2b$$
$$\implies 4a^2b - a^2b - 2ab^2 + 7ab^2 - 3a^2b^2 - 4a^2b^2$$
$$\implies 3a^2b + 5ab^2 - 7a^2b^2$$

Example 1.18

Multiply $7(2x - 3)$. (Use distributive property.)

Solution:
$$7(2x - 3)$$
$$\Longrightarrow \quad 7 \cdot 2x - 7 \cdot 3$$
$$\Longrightarrow \quad 14x - 21$$

Example 1.19

Multiply $\frac{2}{3}\left(- 6x + 27y - 4 \right)$

Solution:
$$\frac{2}{3}\left(- 6x + 27y - 4 \right)$$
$$\Longrightarrow \quad \frac{2}{3} \cdot (-6x) + \frac{2}{3} \cdot 27y - \frac{2}{3} \cdot 4$$
$$\Longrightarrow \quad - 4x + 18y - \frac{8}{3}$$

Example 1.20

Simplify the expression $2x\left(- 2x + 4y + 7 \right) - 3\left(x^2 - 3x\right)$.

Solution:
$$2x\left(- 2x + 4y + 7 \right) - 3\left(x^2 - 3x\right)$$
$$\Longrightarrow \quad 2x\left(- 2x + 4y + 7 \right) + (-3)\left(x^2 - 3x\right)$$
$$\Longrightarrow \quad (2x)(-2x) + (2x)(4y) + (2x)(7) + (-3)\left(x^2\right) - (-3)(3x)$$
$$\Longrightarrow \quad - 4x^2 + 8xy + 14x - 3x^2 + 9x$$

$$\Longrightarrow \quad \underbrace{-4x^2 - 3x^2}_{\text{like terms}} + 8xy + \underbrace{14x + 9x}_{\text{like terms}}$$
$$\Longrightarrow \quad - 7x^2 + 8xy + 23x$$

There is one particular kind of expression simplification that is extremely important to understand, simplifying polynomials that are raised to powers. The most frequently encountered example is squaring a binomial. Most students know this as FOILing. For example, if we want to simplify $(2x+3)^2$, we would write it as $(2x + 3)(2x + 3)$ and then multiply the two "First" terms, the two "Outer" terms, the two "Inner" terms, and the two "Last" terms, and then sum:

$$\begin{array}{rl}
\text{First:} & (2x)(2x) = 4x^2 \\
\text{Outer:} & (2x)(3) = 6x \\
\text{Inner:} & (3)(2x) = 6x \\
\text{Last:} & (3)(3) = 9 \\
\hline
\text{Sum:} & 4x^2 + 12x + 9
\end{array}$$

There is one drawback, FOILing only works when you are squaring a binomial. In more general situations it fails. However, using the distributive property repeatedly always works. Though it is more work, it does allow you to raise binomials to higher powers, or raise polynomials with more than two terms to some power.

Example 1.21

Simplify the expression $(2x + 3)^2$ by FOILing.

Solution:

$$
\begin{aligned}
& (2x + 3)^2 \\
\implies & (2x + 3)(2x + 3) \\
\implies & 4x^2 + 6x + 6x + 9 \\
\implies & 4x^2 + 12x + 9
\end{aligned}
$$

Example 1.22

Simplify the expression $(2x + 3)^2$ using the distributive property.

Solution:

$$
\begin{aligned}
& (2x + 3)^2 \\
\implies & (2x + 3)(2x + 3) \\
\implies & (2x + 3)2x + (2x + 3)3 \quad \text{(Distributive Property)} \\
\implies & \Big(2x(2x) + 3(2x)\Big) + \Big(2x(3) + 3(3)\Big) \quad \text{(Distributive Property)} \\
\implies & (4x^2 + 6x) + (6x + 9) \\
\implies & 4x^2 + 12x + 9
\end{aligned}
$$

Example 1.23

Simplify the expression $(2x + 3)^3$ using the distributive property.

Solution:

$$(2x + 3)^3$$
$$\implies (2x + 3)(2x + 3)(2x + 3)$$
$$\implies (4x^2 + 12x + 9)(2x + 3) \quad \text{(Result from last example.)}$$
$$\implies (4x^2 + 12x + 9)2x + (4x^2 + 12x + 9)3 \quad \text{(Dist. Prop.)}$$
$$\implies 4x^2(2x) + 12x(2x) + 9(2x) + 4x^2(3) + 12x(3) + 9(3)$$
$$\text{(Dist. Prop.)}$$
$$\implies 8x^3 + 24x^2 + 18x + 12x^2 + 36x + 27$$
$$\implies 8x^3 + 36x^2 + 54x + 27$$

One of the most important skills with polynomials is factoring. In a sense, it is a lot like distribution, only backwards. In general, much of high school algebra is taken up with learning a large number of techniques for factoring expressions. Here we will only review the basics, restricting ourselves to low-degree polynomials that you are most likely to encounter in standard math problems. Factoring is often a necessary step in solving for values of x in algebraic equations. Consider the following distribution we did earlier:

$$7(2x - 3) = 14x - 21.$$

Had we been given the expression $14x - 21$ instead and wanted to factor it, what would we have done? Notice that $14 = 7 \cdot 2$ and $21 = 7 \cdot 3$. Thus, we could write $14x - 21 = 7 \cdot 2x + 7 \cdot 3$. Each term in the polynomial expression has a 7 in it. We can factor out whatever is common to all terms. Thus, we have

$$14x - 21 = 7 \cdot 2x + 7 \cdot 3$$
$$= 7(2x + 3).$$

We could check our work by distributing the 7.

Example 1.24

Factor the expression $4x^2 - 12x$.

Solution: We notice that there is a $4x$ common to both terms. Thus, we have

$$4x^2 - 12x = 4x \cdot x - 4x \cdot 3 = 4x(x - 3).$$

Solving equations that involve factoring usually requires what is often called the **zero product property**. The zero product property states that if $a \cdot b = 0$ then either $a = 0$ or $b = 0$ or both a and b equal zero. In other words, if the product of two numbers or factors is zero, then at least one of the numbers or factors must be zero. This, of course, applies to products of more than two factors.

Example 1.25

Solve the equation $15x^3 + 25x^2 = 0$.

Solution: Here we notice that $5x^2$ is common to both terms. Thus we can factor the left-hand side of the equation to get

$$15x^3 + 25x^2 = 5x^2(3x + 5) = 0.$$

This is when zero product property is used. The left-hand side of the equation is $5 \cdot x^2 \cdot (3x+5)$. Clearly, 5 can never be equal to zero. So, the only way the left-hand side of this equation can equal zero is if either the factor x^2 equals zero or the factor $3x + 5$ equals zero. Thus, we have the two possibilities,

$$x^2 = 0 \quad \text{or} \quad 3x + 5 = 0$$
$$\implies \quad x = 0 \quad \text{or} \quad x = \frac{-5}{3}.$$

Most factoring problems are more complicated than simply factoring out whatever is common to all terms. Consider the following product:

$$\begin{aligned}(a + b)(a - b) &= a(a - b) + b(a - b) \\ &= a^2 - ab + ba - b^2 \\ &= a^2 - b^2.\end{aligned}$$

This gives us the **difference of squares** formula,

$$a^2 - b^2 = (a + b)(a - b),$$

which is probably easiest to just memorize.

Example 1.26

Factor $x^2 - 16$.

Solution: It is clear that this is the difference of two terms that are squares, since clearly $16 = 4^2$. Thus, we have

$$x^2 - 16 = (x+4)(x-4).$$

Example 1.27

Solve the equation $4x^2 - 49 = 0$.

Solution: The expression on the left-hand side is clearly a difference of squares since $4x^2 = (2x)^2$ and $49 = 7^2$. Thus, we have

$$4x^2 - 49 = 0$$
$$\implies (2x+7)(2x-7) = 0$$
$$\implies 2x + 7 = 0 \quad \text{or} \quad 2x - 7 = 0$$
$$\implies x = \frac{-7}{2} \quad \text{or} \quad x = \frac{7}{2}.$$

Quadratic expressions are quite common in introductory calculus books like this one. A quadratic expression has the form

$$ax^2 + bx + c,$$

where the coefficients a, b, and c are real numbers. We begin by looking at examples where the leading coefficient is one. It should be noted that it is not always possible to factor quadratics, but examples like those in this book are generally contrived so that it is possible. Let us consider the quadratic $x^2 + 8x + 15$. Suppose that this quadratic was factorable like this,

$$x^2 + 8x + 15 = (x + \alpha)(x + \beta),$$

where α and β are real numbers. Multiplying together the right-hand side gives us

$$(x + \alpha)(x + \beta) = x^2 + (\alpha + \beta)x + \alpha\beta.$$

Thus, we need to find values of α and β such that

$$\alpha + \beta = 8,$$
$$\alpha\beta = 15.$$

Generally, we need to solve this by trial and error. We want to find an α and a β that when multiplied together give us 15 and when added together give us 15. It is fairly straight forward to see that 3 times 5 is 15 and 3 plus 5 is 8. Thus, we have

$$x^2 + 8x + 15 = (x + 3)(x + 5).$$

Suppose, for a moment, that the solution was not quite so obvious. We could still attempt to find the solution by considering all the ways we could factor 15. For small numbers this is usually possible.

$$1 \cdot 15 = 15, \quad \text{or} \quad 3 \cdot 5 = 15.$$

Indeed, our answer is among those two choices.

Example 1.28

Factor the expression $x^2 - 3x - 40$.

Solution: We are looking for two numbers α and β whose sum is -3 and whose product is -40. This is not so immediately obvious, so we consider the factorizations of -40;

$$
\begin{array}{lll}
+1 \cdot -40 & \text{and} & -1 \cdot +40, \\
+2 \cdot -20 & \text{and} & -2 \cdot +20, \\
+4 \cdot -10 & \text{and} & -4 \cdot +10, \\
+8 \cdot -5 & \text{and} & -8 \cdot +5.
\end{array}
$$

Since $-8 + 5 = -3$, we have

$$x^2 - 3x - 40 = (x - 8)(x + 5).$$

Example 1.29

Solve the equation $x^2 + 6x + 8 = 0$.

Solution: Since $2 \cdot 4 = 8$ and $2 + 4 = 6$, we have that

$$
\begin{aligned}
x^2 + 6x + 8 &= 0 \\
\Longrightarrow (x + 2)(x + 4) &= 0 \\
\Longrightarrow x + 2 = 0 \quad &\text{or} \quad x + 4 = 0 \\
\Longrightarrow x = -2 \quad &\text{or} \quad x = -4.
\end{aligned}
$$

The procedure for quadratic equations where the leading coefficient is not one is similar, but with a few additional steps. We will consider the example $ax^2 + bx + c = 2x^2 - x - 6$. In other words $a = 2$, $b = -1$, and $c = -6$. The first step is to multiply a and c to get $a \cdot c = 2 \cdot (-6) = -12$. We are now looking for α and β such that

$$\alpha + \beta = b = -1,$$
$$\alpha\beta = ab = -12.$$

As before, it may be helpful to look at the factorization of -12,

$$+1 \cdot -12 \quad \text{and} \quad -1 \cdot +12,$$
$$+2 \cdot -6 \quad \text{and} \quad -2 \cdot +6,$$
$$+3 \cdot -4 \quad \text{and} \quad -3 \cdot +4.$$

Since $3 + (-4) = -1$ and $3 \cdot (-4) = -12$ then that is the correct option. We then rewrite the quadratic expressions as follows:

$$2x^2 - x - 6 = 2x^2 + (3 - 4)x - 6 = 2x^2 + 3x - 4x - 6.$$

This expression has four terms. We consider the first two terms separately from the lasts two terms, and factor out what is common. Thus, we get

$$2x^2 + 3x - 4x - 6$$
$$\Longrightarrow (2x^2 + 3x) + (-4x - 6)$$
$$\Longrightarrow x(2x + 3) - 2(2x + 3)$$
$$\Longrightarrow (x - 2)(2x + 3).$$

This is the factorization of our quadratic expression. One can, and should, check one's work. It should be noted that the order in which we write the middle terms does not matter in the end. For example, had we written

$$2x^2 - x - 6 = 2x^2 + (-4 + 3)x - 6 = 2x^2 - 4x + 3x - 6$$

then we would have obtained

$$2x^2 - 4x + 3x - 6$$
$$\Longrightarrow (2x^2 - 4x) + (3x - 6)$$
$$\Longrightarrow 2x(x - 2) + 3(x - 2)$$
$$\Longrightarrow (2x + 3)(x - 2).$$

Example 1.30

Factor the expression $4x^2 - 15x + 9$.

Solution: First we find $4 \cdot 9 = 36$. Next, we want to find α and β such that their sum is -15 and their product is 36. This forces us to have both α and β negative,

$$-1 \cdot -36,$$
$$-2 \cdot -18,$$
$$-3 \cdot -12.$$

We could of course continue with this list, but since $-3 - 12 = -15$, then we have found what we needed. Next, we have

$$4x^2 - 15x + 9$$
$$\implies 4x^2 - 3x - 12x + 9$$
$$\implies (4x^2 - 3x) + (-12x + 9)$$
$$\implies x(4x - 3) - 3(4x - 3)$$
$$\implies (x - 3)(4x - 3).$$

Example 1.31

Solve the equation $5x^2 + 7x - 6 = 0$.

Solution: We first find $5 \cdot (-6) = -30$. Next, we consider

$$+1 \cdot -30 \quad \text{and} \quad -1 \cdot +30,$$
$$+2 \cdot -15 \quad \text{and} \quad -2 \cdot +15,$$
$$+3 \cdot -10 \quad \text{and} \quad -3 \cdot +10.$$

Again we could continue with this list, but since $-3 + 10 = 7$, we have found what we need. Next,

$$5x^2 + 7x - 6 = 0$$
$$\implies 5x^2 - 3x + 10x - 6 = 0$$
$$\implies x(5x - 3) + 2(5x - 3) = 0$$
$$\implies (x + 2)(5x - 3) = 0$$
$$\implies x + 2 = 0 \quad \text{or} \quad 5x - 3 = 0$$
$$\implies x = -2 \quad \text{or} \quad x = \frac{3}{5}.$$

If we are asked to solve a quadratic equation

$$ax^2 + bx + c = 0$$

that does not factor easily, we could simply avail ourselves of the **quadratic formula**

$$x_1 = \frac{-b + \sqrt{b^2 - 4ac}}{2a} \quad \text{and} \quad x_2 = \frac{-b - \sqrt{b^2 - 4ac}}{2a},$$

which is, of course, usually written as

$$x = \frac{-b \pm \sqrt{b^2 - 4ac}}{2a}.$$

The quadratic formula is usually taught in high school algebra classes and we will not attempt to prove it here.

Example 1.32

Solve the equation $x^2 - 7x - 3 = 0$.

Solution: A few moments of work will convince you that this quadratic expression does not factor nicely. Thus, the only way we have to solve this equation is to use the quadratic formula,

$$x = \frac{-b \pm \sqrt{b^2 - 4ac}}{2a}$$
$$= \frac{-(-7) \pm \sqrt{(-7)^2 - 4(1)(-3)}}{2(1)}$$
$$= \frac{7 \pm \sqrt{49 + 12}}{2}$$
$$= \frac{7 \pm \sqrt{61}}{2}$$
$$\approx 7.40512, -0.40512.$$

It is usually a good idea to program the quadratic formula into your calculator.

There are, of course, more techniques that can be used for factoring polynomials, but the techniques just covered are sufficient to tackle most examples encountered in this book. But you should beware, the examples in this book are usually contrived to allow you to factor them. Examples and equations you encounter in real life situations are not so nice. But the purpose of this book is to teach you calculus and calculus related concepts. In order to allow us to focus on the calculus we choose examples where the algebra is reasonably nice.

1.4 FUNCTIONS AND GRAPHS

To solve business and economics problems, it is necessary to have a good grasp of functions since they are used to represent relationships between the different economic and business variables. We begin this section by defining functions and then spend some time presenting some of the most important elementary function and their graphs.

Suppose X and Y are sets. A **function** f from set X to set Y, written as $f : X \to Y$, is a rule that assigns to each element x of X exactly one element y of Y. Often we would write this rule as $y = f(x)$ or $x \mapsto y$.

If the rule f sends x in X to y in Y then x is called the **input** variable and y is said to be the **output** variable of f or the **image** of x under f. Usually, x is considered the **independent variable**, while y is considered the **dependent variable**. In general, we think of the value y as a value that depends on what the input value x is.

The most common set used for X and Y in calculus, business, and economics is the set of real numbers. Thus, most functions we will be concerned with send real numbers to real numbers; $f : \mathbb{R} \to \mathbb{R}$. Also, the functions we will deal with in this book are generally written as algebraic expressions.

Example 1.33

Suppose we have the function $f : \mathbb{R} \to \mathbb{R}$ given by $f(x) = x^2 - 2$. Evaluate the function at the following values: 0, -2, 3, h, and $x + 1$.

Solution: To evaluate the function at a particular value, we simply substitute the value of x into the algebraic expression. Thus, we have $f(0) = -2$, $f(-2) = 2$, and $f(3) = 7$.

If the value is simply another variable, we treat it exactly in the same way. $f(h) = h^2 - 2$ and $f(x+1) = (x+1)^2 - 2 = x^2 + 2x + 1 - 2 = x^2 + 2x - 1$.

We define the addition, difference, product, and quotient of two functions as follows:

$$(f + g)(x) = f(x) + g(x)$$
$$(f - g)(x) = f(x) - g(x)$$
$$(fg)(x) = \big(f(x)\big)\big(g(x)\big)$$
$$\left(\frac{f}{g}\right)(x) = \frac{f(x)}{g(x)}, \text{ when } g(x) \neq 0$$

We can also define the **composition** of two functions $y = g(z)$ and $z = f(x)$ as $(g \circ f)(x) = g(f(x))$. This new function $g \circ f$ can be read as "g composed with f" or "g circle f" and denoted:

$$g \circ f : x \xrightarrow{f} z \xrightarrow{g} y = g\big(f(x)\big)$$

$$g \circ f : x \xrightarrow{g \circ f} y = g\big(f(x)\big)$$

It is important to note that in general $g \circ f \neq f \circ g$.

Example 1.34

Given $f(x) = x^2 - 2$ and $g(x) = 3x + 1$. Find $f \circ g$ and $g \circ f$.

Solution:
$$f \circ g(x) = f(g(x)) = f(3x + 1)$$
$$= (3x + 1)^2 - 2$$
$$= 9x^2 + 6x + 1 - 2$$
$$= 9x^2 + 6x - 1$$

and
$$g \circ f(x) = g(f(x)) = g(x^2 - 2)$$
$$= 3(x^2 - 2) + 1$$
$$= 3x^2 - 6 + 1$$
$$= 3x^2 - 5$$

The function $f : X \to Y$ is said to be **onto** Y if, for each element y of Y, there exists $x \in X$ such that $y = f(x)$. The function f is said to be **one-to-one** if for any two elements x_1 and x_2 of X, if $f(x_1) = f(x_2)$ then we must have $x_1 = x_2$. In other words, if two elements in the domain give the same image, those two elements must in fact be the same element.

Example 1.35

Suppose $X = (0, \infty)$, $Y = \mathbb{R}$, and the function $f : X \to Y$ defined by $f(x) = \sqrt{x} + 2$. Is f onto or one-to-one?

Solution: For y any real number, if f is onto $\mathbb{R}$ then there must exist an $x \in (0, \infty)$ such that $y = \sqrt{x} + 2$. Now suppose we choose $y = 0$, by solving the equation $y = \sqrt{x} + 2$ for this y value we get $0 = \sqrt{x} + 2 \implies \sqrt{x} = -2$. But since $\sqrt{x}$ must be a non-negative number then this equation has no solution. Thus f is not onto $\mathbb{R}$.

Now consider x_1 and x_2 two elements of $(0, \infty)$ such that $f(x_1) = f(x_2)$. We have:

$$f(x_1) = f(x_2) \implies \sqrt{x_1} + 2 = \sqrt{x_2} + 2$$
$$\implies \sqrt{x_1} = \sqrt{x_2} \quad \text{(square both sides of the equation)}$$
$$\implies x_1 = x_2 \quad \text{(since } x_1 \text{ and } x_2 \text{ are both positive)}$$

Hence, the function f is one-to-one.

If $f : X \to Y; y = f(x)$ is a one-to-one function, there exists a **inverse function** denoted f^{-1} such that $f \circ f^{-1}(x) = x$ for any $x \in X$. We find $f^{-1}(x)$

by solving the equation $y = f(x)$ with respect to x, which gives $x = f^{-1}(y)$ that is to say that y is the independent variable and x the dependent variable. Then switch x and y to write the inverse function in the conventional form $y = f^{-1}(x)$.

Example 1.36

Given $f : \mathbb{R} \to \mathbb{R}$, with $f(x) = 3x + 5$. Find the inverse of f.

Solution:

$$y = f(x) \Rightarrow y = 3x + 5$$
$$\Longrightarrow \quad 3x = y - 5 \qquad \text{(Solve for } x\text{)}$$
$$\Longrightarrow \quad x = \frac{y - 5}{3}$$
$$\Longrightarrow \quad y = \frac{x - 5}{3} \qquad \text{(Switch } x \text{ and } y\text{)}$$

Therefore $f^{-1}(x) = \frac{x-5}{3}$.

Usually, functions are written in **explicit form** $y = f(x)$ but sometimes they may be written in an **implicit form,** which means that x and y will be on the same side of the equation. Here $y = 3x^2 + 2x - 5$ is an explicit function, whereas $x^2 + 2y = 9$ is an implicit function. It is often, though not always, possible to solve the implicit function and write it in an explicit form.

Now we will consider some properties of functions and explore some elementary functions. We will define both the domain and the range of a function as follows. The **domain** of the function $f : X \to Y$, which we will denote by D_f, is the set of elements $x \in X$ such that $f(x)$ exists. The **range** of the function $f : X \to Y$, which we will denote by R_f, is the set of values $y \in Y$ such that $f(x)$ exists. Another way to say this is that y is the image of an element x of D_f.

Some functions are defined on the set of all real numbers $\mathbb{R}$, while some are defined only on proper subsets of $\mathbb{R}$. Often the domain of a function is restricted by the type of application in which the function arises. For example, if $R(x)$ is the revenue function from selling x products, x is naturally a positive number. Therefore, $D_f = \{x \in \mathbb{R} \mid x \geq 0\} = [0, \infty) = \mathbb{R}^+$. So in general, we have $D_f \subseteq X$ and $R_y \subseteq Y$.

Example 1.37

The function $f(x) = 3x + 1$ is defined on all real numbers since x can be any real number, so $D_f = \mathbb{R}$. It is also easy to see that $y = 3x + 1$ can also output any real number. Therefore, the range of the function f is all real numbers $R_f = \mathbb{R}$

The function $g(x) = \sqrt{x} + 1$ is defined only on zero or positive numbers because of the square-root, therefore, $D_f = \mathbb{R}^+$. Since $\sqrt{x}$ must be 0 or positive, the range is all numbers larger than or equal to 1; $R_f = [1, \infty)$.

Example 1.38

The function $f(x) = \frac{x^2+1}{x-2}$ is defined as long as the denominator is not equal to zero, in other words, as long as $x \neq 2$. Thus we would write $D_f = (-\infty, 2) \cup (2, \infty)$. The $\cup$ mean union. In other words, we are "combining" these two intervals to make one set. We could also write $D_f = \{x \in \mathbb{R} \mid x \neq 2\}$.

The first elementary functions we will be looking at are the linear functions written as $y = f(x) = ax + b$, where a and b are two real numbers. These functions are called linear functions because their graphs are straight lines. In $y = ax + b$, a is called the slope (or gradient) of the line, and b is called the y-intercept. In fact, $x = 0 \implies y = a(0) + b = b$. So $(0, b)$ is a point on the graph of the linear function $y = ax + b$.

To graph a linear function, it is enough to have two points to represent the line, $(0, b)$ could be used as one of the points. Another way to graph it is to use the y-intercept and the slope of the line.

Example 1.39

Graph $y = 2x - 3$.

Solution: If $x = 0$ then $y = -3$, therefore we use $A = (0, -3)$ as first point of the line. To get a second point we choose another value for x, such as $x = 2$. In this case: $y = 2(2) - 3 = 1$, and we have the second point $B = (2, 1)$. We plot the two points and then connect them with a straight line.

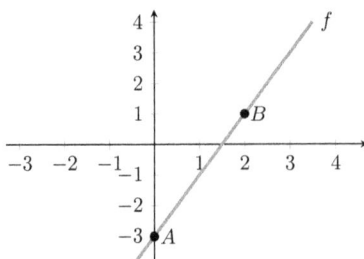

In $y = ax + b$, if $a = 0$, then we have $y = b$, and the graph is a horizontal line at a "height" of b. Similarly, if $b = 0$, then we have $y = ax$, which passes through the origin $(0,0)$. It is also important to note that vertical lines differ from linear functions. A vertical line has the equation of $x = a$, which has a line as graph but is not a function since as for the unique value of $x = 4$, y can take any value.

Example 1.40

The graph of the linear function $f(x) = 2$ is a horizontal line that goes through $y = 2$ for any value of x.

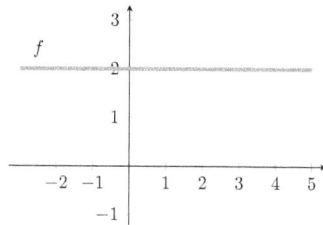

The graphs of two linear functions are parallel if they have same slope as it can be seen in the example below

---

Example 1.41

The graphs of the linear functions $f(x) = 2x - 1$ and $g(x) = 2x + 3$ are parallel since they have the same slope $a = 2$.

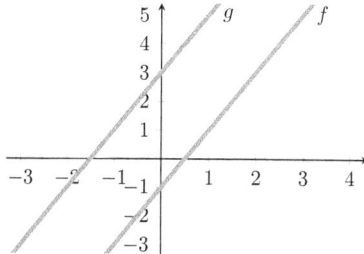

Polynomial functions in one variable are functions of one variable that can be expressed using a polynomial,

$$y = f(x) = a_n x^n + a_{n-1} x^{n-1} + \ldots + a_1 x + a_0,$$

where $n \in \mathbb{N}$, $a_n \neq 0$, and $a_0, a_1, \ldots, a_n \in \mathbb{R}$. Polynomials written this way are said to be in **standard form**. The **degree** of the polynomial is given by the largest exponent, which is n. Thus, $y = 3x^5 - 2x^3 + 4x - 1$ and $f(x) = 7x^2 - 5x + 3$ are polynomials functions of degrees 5 and 2 respectively. If $n = 1, a_1 = a, a_0 = b$ we obtain the linear function $y = ax + b$.

Consider a polynomial function of degree two written in standard form,

$$f(x) = ax^2 + bx + c$$

where a, b, c are real numbers and $a \neq 0$. The graphs of degree two polynomials are parabolas. Writing degree two polynomials in what is called the **vertex form** allows one to graph these parabolas easily. The vertex form is given by

$$f(x) = a(x - h)^2 + k,$$

where $h = \frac{-b}{2a}$, and $k = f(h)$. (To get k substitute h into the standard form of the function.) The parabola vertex is given by the point (h, k) and the axis of symmetry is the line $x = h$. Thus, h is the horizontal shift of the parabola and k is the vertical shift of the parabola. The sign of a determines if the parabola opens up or down.

Example 1.42

The graph of the functions $f(x) = 2x^2 - 4x + 5 = 2(x - 1)^2 + 3$ is an upward parabola as $a = 2 > 0$. Also, the parabola vertex is given by the point $(1, 3)$.

The graph of the function $g(x) = -2x^2 + 12x - 14 = -2(x - 3)^2 + 4$ is a downward parabola because $a = -2 < 0$, with the parabola vertex given by the point $(3, 4)$.

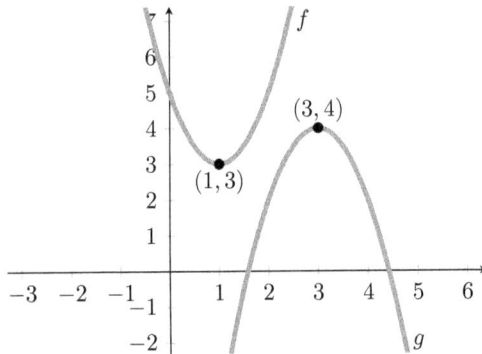

Example 1.43

If $a = 1, b = c = 0$, we have $f(x) = x^2$, which is sometimes called the square function. Since $a = 1 > 0$ the parabola opens upward. The graph of the function $g(x) = -x^2$, is a reflection of the graph of the square function with respect to the x-axis. Here $a = -1 < 0$ and so the parabola opens downward.

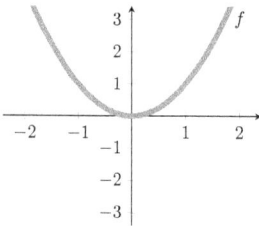

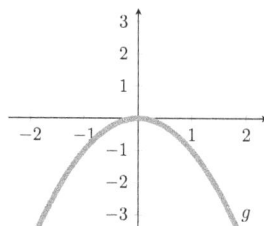

Example 1.44

The below graphs of functions $f(x) = a(x - h)^2 + k$ were obtained as transformations to the square function graph $f(x) = x^2$. Notice how the sign of a determines if the parabola opens up or down, how h shifts the parabola horizontally, and how k shifts the parabola vertically.

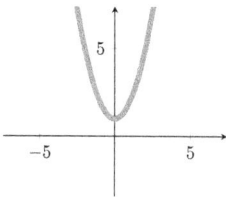

$$f(x) = x^2 + 1$$
$$a = 1, \ h = 0, \ k = 1$$

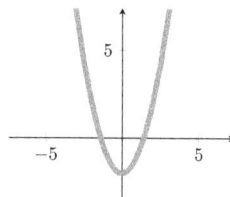

$$f(x) = x^2 - 2$$
$$a = 1, \ h = 0, \ k = -2$$

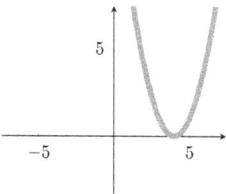

$$f(x) = (x - 4)^2$$
$$a = 1, \ h = 4, \ k = 0$$

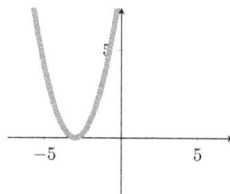

$$f(x) = (x + 3)^2$$
$$a = 1, \ h = -3, \ k = 0$$

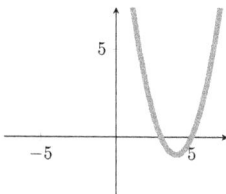

$$f(x) = (x - 4)^2 - 1$$
$$a = 1, \ h = 4, \ k = -1$$

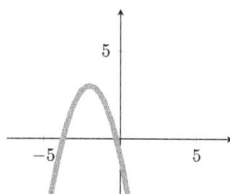

$$f(x) = -(x + 2)^2 + 3$$
$$a = -1, \ h = -2, \ k = 3$$

Another type of functions we will be using are the rational functions, which are written as $y = \frac{p(x)}{q(x)}$, where $p(x)$ and $q(x)$ are two polynomials functions and $q(x) \neq 0$.

Example 1.45

The elementary function $y = \frac{1}{x}$ is one of the examples of rational functions.

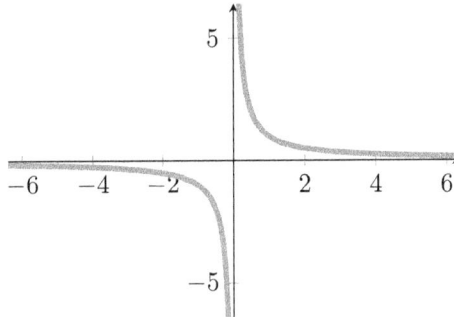

Notice that there is a vertical asymptote at $x = 0$ and the function is undefined at $x = 0$ since division by zero is not allowed. Hence, the inverse function is defined on all real numbers except 0, and we would say the domain is given by $D_f = \mathbb{R} - \{0\}$.

Other types of functions that are essential in mathematics are exponential and logarithmic functions will be dealt with separately in Chapter 3.

1.5 ALGEBRA IN BUSINESS APPLICATIONS

Linear Functions Applications

One of the most studied functions in economics is the **price-demand function**. The price demand function is also often simply called the **demand function**. The demand function is a function that relates consumer demand for a certain quantity of an item with factors that affect the demand. The most important factor is usually price, so in most situations, the demand function is an equation, or relationship, between price and quantity. It is important to remember that the price-demand function is a function that is from the consumer's perspective. The economic principle behind it is that as the price of an item increases, the number of consumers willing to pay that price decreases.

As a first approximation, we assume there is a linear relationship between price and quantity demanded; the more one charges for an item the less consumer demand there will be for the item. Thus, the demand function is a linear function with a negative slope; $y = ax + b$ where $a < 0$. In this function, x represents the quantity demanded while y represents the price p. In business and economics often the variable q is used to represent quantity and the variable p is used to represent price, so we could write the equation as $p = aq + b$. As price and quantity are always positive numbers, we will only use the first quadrant of the Cartesian system when graphing the demand function.

Example 1.46

Suppose $y = -0.2x + 10$, or $p = -0.2q + 10$, is a price-demand function where the slope is -0.2 and the y-intercept is 10. Thus, this graph shows that for each \$0.20 decrease in price, the demand increases by one unit.

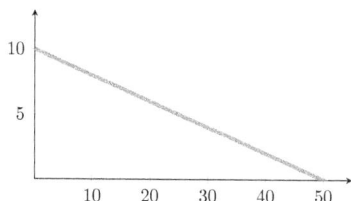

Example 1.47

For a commodity, the demand is 30 if the price is \$30 and is 10 if the price is \$40. Find the demand function.

Solution: As the demand function is a linear function it can be written as $p = ax + b$. We have two points that belong to the line that represent the graph of the demand function, $(30, 30)$ and $(10, 40)$. Thus we can easily calculate the slope of the line:

$$a = \frac{p_2 - p_1}{x_2 - x_1} = \frac{40 - 30}{10 - 30} = \frac{10}{-20} = -\frac{1}{2}$$

So, $p = -\frac{1}{2}x + b$. We can obtain the value b by substituting one of the two points, say $(10, 40)$, into the equation:

$$40 = -\frac{1}{2}(10) + b \quad \Longrightarrow \quad 40 = -5 + b \quad \Longrightarrow \quad b = 45$$

Therefore, the demand function is $p = -\frac{1}{2}x + 45$ or $p = -\frac{1}{2}q + 45$. We can also write this equation as: $2p = -q + 90$ or even $q = 90 - 2p$.

Another important function to study, in economics, is the **supply function**, which is a function that relates the supply of an item with factors that affect the supply. Again, the most important factor is usually price, so in most situations the supply function is a relationship between the quantity and price. Again, as a first approximation, we generally assume there is a linear relationship between price and quantity; the more one can charge for an item the more items are supplied by producers.

Here we should remember that the supply function is a function that is from the producer's perspective. Thus the supply function is a linear functions with positive slope; $y = ax + b$ where $a > 0$. Using q for quantity and p for price we again have $p = aq + b$. Similar to the demand function, price and quantity are positive integers so when graphing the supply function we are only interested in the first quadrant.

Example 1.48

Suppose $y = 0.2x + 5$ or $p = 0.2q + 5$ is a supply function where the slope is 0.2 and the y-intercept is 5. Thus, this graph shows that for each \$0.20 increase in price, the supply (or production) increases by one unit.

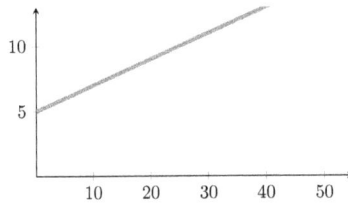

Example 1.49

When the price of a commodity is \$30, the supplied quantity is 130, and when the price is \$15, the supply is 100. Find the supply function.

Solution: As the supply function is a linear function it can be written as $p = ax + b$. We have two points that belongs to the line that represent the graph of the demand function, $(130, 30)$ and $(100, 15)$. The slope of the line can be calculated easily as:

$$a = \frac{p_2 - p_1}{x_2 - x_1} = \frac{30 - 15}{130 - 100} = \frac{15}{30} = \frac{1}{2}$$

So, $y = \frac{1}{2}x + b$. We can obtain the value b by substituting one of the two points, say $(130, 30)$, into the equation:

$$30 = \frac{1}{2}(130) + b \quad \Longrightarrow \quad 30 = 65 + b \quad \Longrightarrow \quad b = -35$$

Therefore, the supply function is $y = \frac{1}{2}x - 35$ or $p = \frac{1}{2}q - 35$. We can also write this equation as $2p = q - 70$ or even $q = 2p + 70$.

The demand function relates the demand for an item with the price of the item, and the supply function relates the supply of an item with the price of the item. So how many items are actually produced? We define what is called the **equilibrium point** (EP) as the point where the demanded quantity is equal to the supplied quantity. In other words, the EP point occurs when demand = supply.

Example 1.50

Suppose for some commodity we had a demand function $p = -0.2q + 10$ and a supply function $p = 0.2q + 5$. What quantity of the item is produced?

Solution: The quantity of the item produced occurs at the equilibrium point, which occurs when demand = supply. One way to find the equilibrium point is graphically; we graph both the demand function and the supply function and find their point of intersection.

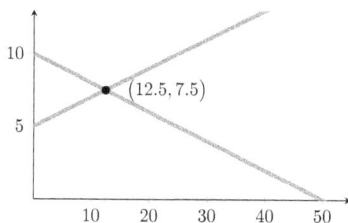

Since in the real world half an item is not produced, we round to the nearest whole number, which is 13. Thus, 13 items would be produced.

Example 1.51

Find the EP for the demand function $y = -x + 6$ and the supply function $y = \frac{1}{3}x + 2$.

Solution:
$$\text{Demand} = \text{Supply}$$
$$\implies \quad -x + 6 = \frac{x}{3} + 2$$
$$\implies \quad 3(-x + 6) = 3\left(\frac{x}{3} + 2\right)$$
$$\implies \quad -3x + 18 = x + 6$$
$$\implies \quad 4x = 12$$
$$\implies \quad x = 3$$

Then we substitute the quantity $x = 3$ into one of the equations to find that price $p = 3$.

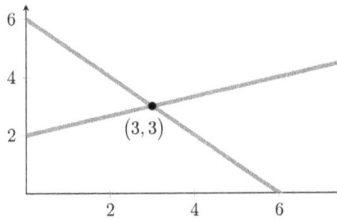

Another important function is the **consumption function** written as $C = f(R)$, where C represents the consumption and R represents the revenue. When revenue $R = 0$, consumption C cannot be zero as there is always a minimum vital consumption even without any revenue. When revenue increases, consumption increases, so it can represented by a linear function with a positive slope $a > 0$. As an example, $C = 0.2R + 10$ is a linear consumption function.

More Applications of Functions in Business and Economics

Thus far we have seen linear demand and supply functions. But it is also possible for these two functions to be non-linear. For example, we could have $y = x^2 + 3x + 1$ as a supply function and $y = -x^2 + 5$ as a demand function. As we have seen before, demand has a negative slope, in other words, when the price increases, demand decreases. Supply has a positive slope which means that if the quantity supplied increases, the price decreases.

Example 1.52

Find the EP when the demand function for widgets is $p = x^2 + 3x + 1$ and the supply function for widgets is $p = -x^2 + 10$, where quantity x is in thousands of widgets. How many widgets will be produced and what will the cost be?

Solution:

$$\text{Demand} = \text{Supply}$$
$$\implies x^2 + 3x + 1 = -x^2 + 10$$
$$\implies 2x^2 + 3x - 9 = 0$$
$$\implies x = \frac{-3 \pm \sqrt{3^2 - 4(2)(-9)}}{2(2)}$$
$$\implies x = \frac{3}{2} = 1.5 \quad \text{or} \quad x = -3$$

Here we used the quadratic formula $x = \frac{-b \pm \sqrt{b^2 - 4ac}}{2a}$ to find two values of x. Since x must be positive, we have $x = 1.5$. Substituting x back into one of the equations we obtain the price $y = -(1.5)^2 + 10 = 7.75$. Thus the equilibrium point is $(1.5, 7.75)$. Since x is given in thousands the equilibrium point tells us that 1500 widgets are produced and are sold for \$7.75 per widget. Graphically we have:

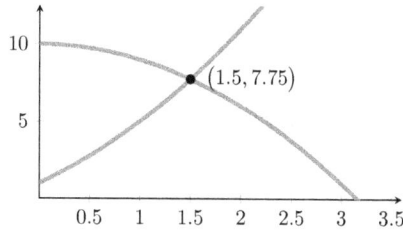

Example 1.53

Find the EP for the demand function $y = 2x + 1$ and the supply function $y = \frac{5}{x+2} + 1$, where the quantity is given in millions.

Solution:
$$\text{Demand} = \text{Supply}$$
$$\implies 2x + 1 = \frac{5}{x+2} + 1$$
$$\implies (x+2)(2x+1) = 5 + (x+2)$$
$$\implies 2x^2 + 4x - 5 = 0$$
$$\implies x = \frac{-4 \pm \sqrt{4^2 - 4(2)(-5)}}{2(2)}$$
$$\implies x = -2.871 \text{ or } x = 0.871$$

The quantity must be positive so $x = 0.871$ is the solution. We find the price by substituting x into one of the equations, $y = 2(0.871) + 1 = 2.74$. Recalling that quantity is given in millions, production is 871,000 with a per unit price of \$2.74. Graphically we have:

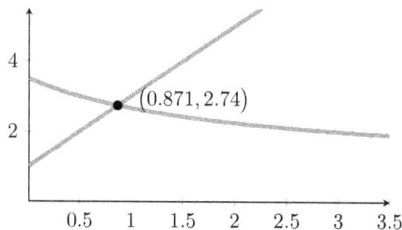

Cost Functions

There is, of course, a cost associated with the production of any item. In the simplest scenario, there are two different kinds of costs, **fixed costs** and **variable costs**.

- Fixed costs, denoted FC, are the costs that are fixed irrespective of the output. This might include the rent for the building, which must be paid no matter how much production there is.
- Variable costs, denoted VC, are the costs that vary with the output. This usually includes things like the cost of the materials used in the production, since how much materials you buy depends on how much you produce. Thus, variable costs are related to production levels.

The **total cost function**, denoted TC or $C(x)$, where x is the number of units output, is the addition of both fixed costs and variables costs, $TC = C(x) = FC + VC$.

Example 1.54

A small artisan bread baking shop has fixed costs of $40 for the rent of a stall in the town market and variable costs of $3 per loaf of bread. Write an equation describing the total cost function of the baking shop and graph it. Find the total costs of producing 32 loaves of bread.

Solution: We begin by denoting the number of loaves of bread by x. Since variable costs are $3 per loaf of bread, we can write $VC = 3x$. Fixed costs are simply written as $FC = 40$. Adding together gives us total cost, $TC = FC + VC = 40 + 3x$, or $C(x) = 40 + 3x$.

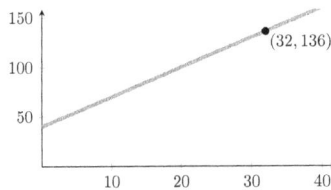

Producing 32 loaves of bread means $x = 32$. By substituting into the TC equation, we find $TC = C(32) = 40 + 3(32) = \$136$. Thus the total cost of producing 32 loaves of bread is $136.

Revenue Functions

A firm receives revenue when it sells output. The **total revenue function**, denoted TR or $R(x)$, is the price of the product multiplied by the number of units sold. Letting p be the price of the product and x the number of units sold, we have $TR = R(x) = px$.

Example 1.55

A small artisan bread baking shop sells bread for $6 a loaf. Write an equation for the total revenue function and graph it. Find the baking shop revenue from selling 32 loaves of bread.

Solution: Letting x represent the number of loaves of bread sold, the total revenue function is given by $R(x) = px = 6x$.

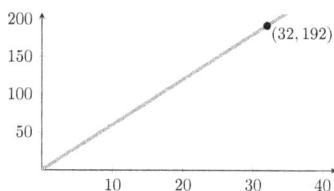

Selling 32 loaves of bread results in a revenue of $R(32) = 6(32) = 192$. Thus the total revenue from selling 32 loaves of bread is $192.

Profit Functions

Profit is the total revenue received minus the total cost. Thus the **profit function**, denoted by $P(x)$, is giving by the formula $P(x) = TR - TC = R(x) - C(x)$. The profit is also sometimes called the **total net revenue** or just the **net revenue**. Net revenue means the revenue obtained after costs are deducted.

Example 1.56

Find the profit function of the small artisan bread baking shop presented in examples 1.54 and 1.55 and graph it. What is the baking shop profit from selling 32 loaves of bread?

Solution:

$$P(x) = R(x) - C(x)$$
$$\implies P(x) = 6x - (40 + 3x)$$
$$\implies P(x) = 6x - 40 - 3x$$
$$\implies P(x) = 3x - 40$$

Notice when we graph the profit function we allow negative numbers. A negative profit is usually called a **loss**.

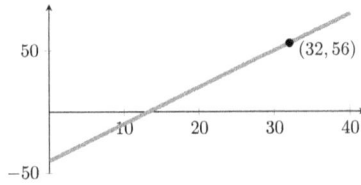

Selling 32 loaves of bread results in a total profit of $P(32) = 56$.

The **break-even point** for a commodity happens when total revenue is equal to total costs. This is the same as saying that the break-even point occurs when profit is equal to zero.

Example 1.57

Find the break-even point for the artisan bread shop in examples 1.54 - 1.56.

Solution: There are several ways we could do this. First, we could take the profit function from example 1.56, set it equal to zero, and solve for x.

$$P(x) = 0$$
$$\implies 3x - 40 = 0$$
$$\implies 3x = 40$$
$$\implies x = \frac{40}{3} \approx 13.3$$

Graphically, the break-even point occurs when the graph of the profit function crosses the x-axis.

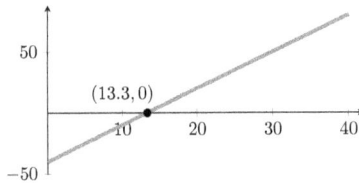

A second approach is to take the cost function from example 1.54 and set it equal to the revenue function from example 1.55, and then solve for x.

$$C(x) = R(x)$$
$$\implies 40 + 3x = 6x$$
$$\implies 40 = 3x$$
$$\implies x = \frac{40}{3} \approx 13.3$$

Which is clearly the same answer already obtained. Graphically,

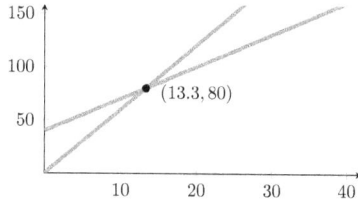

Of course, we would round the final answer to the nearest whole number, so we would say the break-even point occurs when 13 loaves of bread are sold.

Example 1.58

The total revenue and total costs functions for quantity x are given as:

$$TR = R(x) = 4x$$
$$TC = C(x) = 12 + 3x$$

1. Find the equilibrium quantity algebraically at the break-even-point.
2. Find the total revenue and total costs at the break-even-point.

Solution:

1. To find the break-even point, we need to equate TR and TC

$$TR = TC$$
$$\Rightarrow \quad 4x = 12 + 3x$$
$$\Rightarrow \quad 4x - 3x = 12$$
$$\Rightarrow \quad x = 12$$

So, the equilibrium quantity is at the break-even-point, which is 12.

2. To find the total revenue and total costs at the break-even-point, it is enough to substitute the quantity $x = 12$ in their respective equations.

$$TR = 4(12) = 48,$$
$$TC = 12 + 3(12) = 48.$$

At $x = 12$, $TR = TC = 48$.

1.6 PROBLEMS

Question 1.1 *Indicate whether each of the following statements is true or false:*

(a) *5 is a natural number.*

(b) *A rational number is a natural number.*

(c) *Each natural number is an integer.*

Question 1.2 *Give an example of a rational number that is not a natural number.*

Question 1.3 *Perform the indicated operation and simplify:*

(a) $(3x - 2)(4 - x)$

(b) $3x^2 - 5[(x - 2)(3 - x) - 1]$

(c) $6 - 2x(x^3 - 2x + 1) + 2x$

(d) $(5-p)p+3p-20(q-4p^2-5p+1)$

(e) $(3x + 5)^3$

(f) $(2x^2 - x + 3)^2$

Question 1.4 *Given $f(x) = x^2 + 3x - 1$ find $f(h + 2)$.*

Question 1.5 *Given $f(p) = p(p - 10)$ find $f(10)$.*

Question 1.6 *Given $f(x) = \frac{1}{3}x + 2$ and $g(x) = x^2 - 3$. Find $f \circ g(x)$.*

Question 1.7 *Given $f(x) = \frac{x+2}{3x-1}$ and $g(x) = x^3 + 2x - 1$. Find $f \circ g(1)$.*

Question 1.8 *Factor the following expressions:*

(a) $x^2 - 10x$

(b) $x^3 - 3x^2 + 6x$

(c) $x^2 + x - 6$

(d) $x^3 - 5x^2 + 6x$

(e) $x^3 - 27x$

(f) $x^2 + 5x + 4$

(g) $2x^2 - 3x + 1$

(h) $21x^2 - 23x + 6$

(i) $-3x^2 + 14x - 8$

Question 1.9 *Solve the following equations:*

(a) $(3x - 2)(4 - x) = 0$

(b) $(3x - 5)[(x - 2)(3 - x)] = 0$

(c) $x^3 - 5x^2 + 6x = 0$

(d) $x^3 - 27 = 0$

(e) $x^2 + 5x + 4 = 0$

(f) $21x^2 - 23x + 6 = 0$

Question 1.10 *Find the domain of the following functions:*

(a) $x^3 + 5x - 4$

(b) $\frac{1}{x^2-16}$

(c) $\frac{3}{x^2+6}$

(d) $\frac{x-3}{x^2-9}$

(e) $\sqrt{x-4}$

(f) $\sqrt{x^2+1}$

Question 1.11 *Graph the line with slope 3 and y-intercept -1.*

Question 1.12 *Graph the function $y = 3 - x^2$ on the interval $(-10, 10)$.*

Question 1.13 *Graph the function* $y = 2x^2 - 3x + 6$ *on the interval* $[0, 30)$.

Question 1.14 *When the price of a textbook is \$150, none are sold, when the textbook is given away for free, the demand is 80 textbooks. Write the demand function and graph it.*

Question 1.15 *When the price of a cap of special brew coffee is \$9, 18 are sold, when it is on sale at \$6, the demand increases to 30. Write the demand function and graph it.*

Question 1.16 *A bookshop sells coffee cups when the price per cup exceeds \$3. For every \$1 increase in the price, the bookshop increases its supply by 2 cups. Write down the supply equation and graph it.*

Question 1.17 *Find the equilibrium point, price and quantity, of the market with the following demand and supply functions:*
$$D: \quad p = -4x + 12$$
$$S: \quad p = x + 2$$

Question 1.18 *Find the equilibrium price and quantity of the market with the following demand and supply functions:*
$$D: \quad p = -3x + 9$$
$$S: \quad p = 2x + 4$$

Question 1.19 *Find the equilibrium point of the market with the following demand and supply functions:*
$$D: \quad p = -0.5x + 19$$
$$S: \quad p = 0.01x^2 + 5$$

Question 1.20 *The equilibrium point of the market is at the point* $(2, 5)$. *The line representing the demand function passes through* $(0, 9)$. *Find the demand and supply functions, if possible.*

Question 1.21 *The equilibrium point of the market is at the point* $(4, 9)$. *The line representing the demand function has a slope of* -3. *Find the demand and supply functions, if possible.*

Question 1.22 *The equilibrium point of the market is at the point* $(3, y_0)$. *The supply function is given as:* $y = x^2 + 5$ *while the demand function is:* $y = ax^2 + 3x + 6$. *Find* a *and* y_0.

Question 1.23 *Answer the following questions if the demand function is given by* $p = 120 - \frac{2}{3}x$.

(a) *What does the slope* $-\frac{2}{3}$ *mean?*

(b) *What does the p-intercept* 120 *mean in this case?*

(c) *What is the demand when the price is 10?*

(d) *Graph the demand function for* $0 < x < 100$.

(e) *Write the demand function as* $x = D(p)$, *in other terms solve for the quantity* x.

Question 1.24 *Answer the following questions if the supply function is given by* $p = 120 + \frac{1}{2}x$.

(a) *What does the slope* $\frac{1}{2}$ *mean?*

(b) *What does the p-intercept* 120 *mean?*

(c) *What is the supply when the price is $200?*

(d) *Graph the supply function for* $0 < x < 150$.

(e) *Write the supply function as* $x = D(p)$, *in other terms solve for the quantity* x.

Question 1.25 *A firm producing wall lamps has a fixed cost of $2500. If each wall lamp costs $20, write down the total costs equation and graph it on* $0 < x < 120$.

Question 1.26 *A keto restaurant has a fixed cost of $1500 per month. Each keto meal costs $9 to make and is sold at a fixed price of $12. Answer the following questions:*

(a) *Write own the total costs equation and graph it for* $0 < x < 120$.

(b) *What is the cost of producing* 60 *keto meals per week?*

(c) *What is the number of keto meals produced when the variable costs are $972?*

(d) *Write down the equation of the total revenue.*

(e) *What is the number of keto meals produced when the total costs are $1590?*

(f) *Write down the equation of the profit per week.*

(g) *What is the profit when* 800 *keto meals are sold per week?*

Question 1.27 *A firm that makes light suitcases has a total cost function* $TC = 2000 + 5x$.

(a) *If the price of a suitcase is $85, what is the total revenue function?*

(b) *What is the number of suitcases needed to break even?*

(c) *When the firm charges $p for each suitcase. What is the equation for the break-even point in this case?*

(d) *What is the price p charged per suitcase if the break-even-point is* $x = 120$

CHAPTER 2

Rates of Change and the Derivative

Calculus is an area of mathematics that studies how quantities change and uses this description of change to give extra information about the quantities. Calculus gives information about the change in value of the dependent variable, often called y, with the change in value of the independent variable, often called x. This is of prime importance in business and economics.

2.1 RATES OF CHANGE

Since calculus is the study of the relationship between the change in a quantity y with respect to the change in the quantity x, we will begin by looking at **average rates of change**. We will start off by giving the procedure to find the average rate of change.

Procedure to find the average rate of change of the dependent variable y with respect to the independent variable x.

1. Find the change in quantity x.

   a) Find the first value of x and call it x_1.

   b) Find the second value of x and call it x_2.

   c) Find the difference between x_1 and x_2, that is, find $x_2 - x_1$. This is the change in x.

2. Find the change in quantity y.

   a) Find the first value of y and call it y_1.

   b) Find the second value of y and call it y_2.

   c) Find the difference between y_1 and y_2, that is, find $y_2 - y_1$. This is the change in y.

DOI: 10.1201/9781003480235-2

3. Find the average rate of change of y with respect to x by dividing the change in y by the change in x,

$$\frac{y_2 - y_1}{x_2 - x_1}.$$

Does this formula look familiar? It should. We will discuss this formula in more detail in the next section. For now, we want to take a closer look at the phrase **average rate of change of** y **with respect to** x. This phrase consists of three parts, the word **average**, the phrase **rate of change**, and the phrase **of** y **with respect to** x. We will explain each of these in turn.

- **average**: The word average is used to distinguish the above way of finding a rate of change from instantaneous rates of change that we will study soon. Thus, when you see the phrase average rate of change you know to use the procedure above.

- **rate of change**: The word rate is related to the word ratio, which is another way of saying fraction. Thus the rate of change is a fraction where the numerator and the denominator are changes. Recall, that the numerator of a fraction is the "upstairs" part and the denominator of a fraction is the "downstairs" part.

- **of** y **with respect to** x: This is just a fancy way of saying that the change in y is in the numerator and the change in x is in the denominator.

Average rates of change tell us how much the variable y changes, on average if the variable x changes one unit. This average rate of change is valid between the variables x_1 and x_2. It is time to look at a few examples.

Example 2.1

Let $y = 10x^2$. Find the average rate of change of y with respect to x as x changes from 3 to 5.

Solution: Notice, this question actually gives us two values of x when it says that x changes from 3 to 5. This gives us our first value of x and our second value of x. Any actual problem has to tell you in some way what the two values of x are. Following the above procedure, we have:

1. Find the change in quantity x.

    a) The first value of x is $x_1 = 3$.

    b) The second value of x is $x_2 = 5$.

    c) The change in x is $x_2 - x_1 = 5 - 3 = 2$.

2. Find the change in quantity y.

 a) The value y_1 is found by substituting x_1 into the formula for y,

$$y_1 = 10(x_1)^2 = 10(3)^2 = 90.$$

 b) The value y_2 is found by substituting x_2 into the formula for y,

$$y_2 = 10(x_2)^2 = 10(5)^2 = 250.$$

 c) The change in y is $y_2 - y_1 = 250 - 90 = 160.$

3. The average rate of change of y with respect to x is thus given by $\frac{160}{2} = 80$. This means that between the x values of 3 and 5 the variable y changes, on average, 80 for each unit change in x.

Of course, as you do more examples and get more comfortable with finding average rates of change you will not need to write down all of this detail. Once we know what we are doing finding average rates of change is quite easy.

Example 2.2

Let $y = f(x) = 100x - 20x^2$. Find the average rate of change of y with respect to x as x increases from 2 to 5.

Solution: Notice that the problem again gives us two values of x. Also, notice we are using function notation and that the y variable is given by the function $f(x) = 100x - 20x^2$. The rest of the problem follows just like before. Without going into as much detail, we have

$$x_1 = 2,$$
$$x_2 = 5.$$

We use these two values of x to find the associated values of y,

$$y_1 = f(x_1) = f(2) = 100(2) - 20(2)^2 = 200 - 80 = 120,$$
$$y_2 = f(x_2) = f(5) = 100(5) - 20(5)^2 = 500 - 500 = 0.$$

Thus the average rate of change of y with respect to x as x increases from 2 to 5 is given by

$$\frac{y_2 - y_1}{x_2 - x_1} = \frac{0 - 120}{5 - 2} = -40.$$

2.2 INSTANTANEOUS RATES OF CHANGE

Sometimes we want to know how quickly the quantity y is changing at a single value of x. This is called the **instantaneous rate of change** of y with respect to x at the value of x. How would we calculate this? For example, suppose we wanted to know the instantaneous rate of change of $y = 10x^2$ with respect to x at $x = 2$. One idea is to use the average rate of change formula,

$$\frac{\text{change in } y}{\text{change in } x} = \frac{y_2 - y_1}{x_2 - x_1}.$$

But we only have one value of x, namely 2. If we let $x_1 = 2$ and $x_2 = 2$, we would get $y_1 = 40$ and $y_2 = 40$. This would give

$$\frac{y_2 - y_1}{x_2 - x_1} = \frac{40 - 40}{2 - 2} = \frac{0}{0}$$

which is undefined since division by zero is not allowed. Another idea is to let $x_1 = 2$ and choose a value for x_2 that is very close to 2. Doing this would give an estimate for the instantaneous rate of change.

Example 2.3

Let $y = 10x^2$. Estimate the instantaneous rate of change of y with respect to x at the value $x = 2$ by choosing a second value of x close to 2 and then using the formula for the average rate of change.

Solution: Suppose we choose $x_2 = 2.1$. Then we would have $x_1 = 2$ and $x_2 = 2.1$, which would give us $y_1 = 10(2)^2 = 40$ and $y_2 = 10(2.1)^2 = 44.1$. Using the average rate of change formula, we have

$$\frac{y_2 - y_1}{x_2 - x_1} = \frac{44.1 - 40}{2.1 - 2} = \frac{4.1}{0.1} = 41.$$

Thus, we have estimated the instantaneous rate of change of y with respect to x at $x = 2$ to be 41. But what if I wanted a better estimate? We could have chosen $x_2 = 2.01$. Then we would have $x_1 = 2$ and $x_2 = 2.01$, which would give us $y_1 = 10(2)^2 = 40$ and $y_2 = 10(2.01)^2 = 40.401$. Using the average rate of change formula, we have

$$\frac{y_2 - y_1}{x_2 - x_1} = \frac{40.401 - 40}{2.01 - 2} = \frac{0.401}{0.01} = 40.1.$$

Here the estimated instantaneous rate of change of y with respect to x at $x = 2$ is 40.1. Of course, we could then ask for an even better estimate. Suppose we choose $x_2 = 2.0001$. Then we would have $y_2 = 10(2.0001)^2 =$

40.0040001 which would then give us

$$\frac{y_2 - y_1}{x_2 - x_1} = \frac{40.0040001 - 40}{2.0001 - 2} = \frac{0.0040001}{0.0001} = 40.001$$

so the estimated instantaneous rate of change at $x = 2$ is 40.001.

It seems that the closer our second value of x is to 2 the closer the estimated instantaneous rate of change is to 40. Is there a way to get an exact answer without estimating? Yes there is, we have to use a bit of a trick however. We choose our second value of x to be $x_2 = 2 + h$. By doing this we can let h be as small as we want it to be. Usually what we do is let h get closer and closer to zero. That is, we **let h go to zero**. This is written as $h \to 0$. Often we also say that we **take the limit as h goes to zero**.

Example 2.4

Let $y = 10x^2$. Find the instantaneous rate of change of y with respect to x at $x = 2$ by choosing the second value of x to be $2+h$, using the formula for the average rate of change, and then talking the limit as h goes to zero.

Solution: We still have $x_1 = 2$ which gives us $y_1 = 10(2)^2 = 40$. If we let $x_2 = 2 + h$, then we have

$$y_2 = 10(2 + h)^2 = 10(2 + h)(2 + h)$$
$$= 10(4 + 4h + h^2) = 40 + 40h + 10h^2.$$

Now we use the formula for the average rate of change,

$$\frac{y_2 - y_1}{x_2 - x_1} = \frac{\cancel{40} + 40h + 10h^2 - \cancel{40}}{\cancel{2} + h - \cancel{2}} = \frac{40h + 10h^2}{h}$$
$$= \frac{\cancel{h}(40 + 10h)}{\cancel{h}} = 40 + 10h.$$

Notice how we have factored h out of the numerator in order to cancel with the h in the denominator. We now have to take the limit as h goes to zero. In taking the limit as h approaches 0 we simply substitute 0 in for h and calculate the value of the expression.

$$\lim_{h \to 0} \left(40 + 10h\right) = 40 + 10(0) = 40.$$

It was important to cancel the h in the denominator, so we would not have to divide by zero, which is undefined. Thus, the instantaneous rate of change of y with respect to x at $x = 2$ is given by 40. This is an exact answer.

In this book, we will be somewhat imprecise about the concept of limits, which is sufficient for our needs. In more advanced calculus books meant for science majors and engineers, the limit is treated more precisely. But all of our examples are mathematically simple and substitution will work, usually after we have done some algebra to ensure we are not trying to divide by zero.

Example 2.5

Find the limit of the following function when x approaches 0:

$$f(x) = \frac{x^2 - 9}{x - 3}.$$

Solution: Finding the limit as $x \to 0$ simply means substituting 0 in for x,

$$\lim_{x \to 0} f(x) = \lim_{x \to 0} \frac{x^2 - 9}{x - 3} = \frac{0^2 - 9}{0 - 3} = \frac{-9}{-3} = 3$$

Example 2.6

Find the limit of the following function when x approaches 3:

$$f(x) = \frac{x^2 - 9}{x - 3}.$$

Solution: By submitting $x = 3$ in $f(x)$ we will find the undefined fraction $\frac{0}{0}$, which means we cannot simply substitute. However, we can simplify $f(x)$ by factoring the numerator and canceling,

$$f(x) = \frac{x^2 - 9}{x - 3} = \frac{(x - 3)(x + 3)}{(x - 3)} = x + 3.$$

After this simplification, we find that we can indeed simply substitute the value 3 in for x,

$$\lim_{x \to 3} f(x) = \lim_{x \to 3} (x + 3) = 3 + 3 = 6.$$

More advanced books deal with situations where this substitution is not always possible, but for this book it will be, though perhaps some simplification of the function or some canceling will be necessary. Now suppose we wanted to know the instantaneous rate of change of y with respect to x at $x = 3$. We could proceed similarly.

Example 2.7

Let $y = 10x^2$. Find the instantaneous rate of change of y with respect to x at $x = 3$ by choosing the second value of x to be $3+h$, using the formula for the average rate of change, and then taking the limit as h goes to zero.

Solution: We have $x_1 = 3$ which gives us $y_1 = 10(3)^2 = 90$. If we let $x_2 = 3 + h$, then we have

$$y_2 = 10(3+h)^2 = 10(3+h)(3+h)$$
$$= 10(9 + 6h + h^2) = 90 + 60h + 10h^2.$$

Now we use the formula for the average rate of change,

$$\frac{y_2 - y_1}{x_2 - x_1} = \frac{\cancel{90} + 60h + 10h^2 - \cancel{90}}{\cancel{3} + h - \cancel{3}} = \frac{60h + 10h^2}{h}$$
$$= \frac{\cancel{h}(60 + 10h)}{\cancel{h}} = 60 + 10h.$$

We now have to take the limit as h goes to zero. This is written as

$$\lim_{h \to 0} \left(60 + 10h \right) = 60 + 10(0) = 60.$$

The instantaneous rate of change of y with respect to x at $x = 3$ is 60.

If we wanted to know the instantaneous rate of change at some other value of x, we would have to do the same procedure over again. We do not actually want to do this procedure over again every time our value of x changes. So now we use another trick. Last time we used the variable h to represent a very small number. Now we will use the variable x to represent the value of x we want to find the instantaneous rate of change at. In other words, we will say that $x_1 = x$ and $x_2 = x + h$.

Example 2.8

Let $y = 10x^2$. Find the instantaneous rate of change of y with respect to x at a value x. (Here we let the variable x stand for a number.) Do this by letting the first value of x be $x_1 = x$ and the second value of x be $x_2 = x + h$. Use the formula for the average rate of change and then take the limit as h goes to zero.

Solution: If $x_1 = x$ then $y_1 = 10(x_1)^2 = 10x^2$ and if $x_2 = x + h$ then

$$y_2 = 10(x_2)^2 = 10(x+h)^2 = 10(x+h)(x+h)$$
$$= 10(x^2 + 2xh + h^2) = 10x^2 + 20xh + 10h^2.$$

Now we use the formula for the average rate of change,

$$\frac{y_2 - y_1}{x_2 - x_1} = \frac{\cancel{10x^2} + 20xh + 10h^2 - \cancel{10x^2}}{\cancel{x} + h - \cancel{x}} = \frac{20xh + 10h^2}{h}$$

$$= \frac{\cancel{h}(20x + 10h)}{\cancel{h}} = 20x + 10h.$$

We now have to take the limit as h goes to zero. This is written as

$$\lim_{h \to 0} \left(20x + 10h \right) = 20x + 10(0) = 20x.$$

The instantaneous rate of change of y with respect to x at some value x is $20x$.

It is easy to check this answer against the two examples we did above. If we choose x to be 2, then we have $20(2) = 40$, which was what we got as the instantaneous rate of change of y with respect to x at the value 2. If we choose x to be 3, then we have $20(3) = 60$, which was what we got as the instantaneous rate of change of y with respect to x at the value 3.

2.3 THE DERIVATIVE OF A FUNCTION

The concept of a derivative is very closely related to instantaneous rates of change. In essence, the **derivative** of a function is a new function that gives the instantaneous rate of change of the original function at each point.

To understand this consider the last example where we had $y = 10x^2$. We could write this using function notation as $y = f(x)$ where $f(x) = 10x^2$. At the end of the last example, we had obtained the function $20x$. This defines a new function, which we will call $f'(x) = 20x$, which is the derivative of $f(x)$. It is standard practice to write the derivative of function $f(x)$ as $f'(x)$, pronounced "f prime of x." If we wanted to know the instantaneous rate of change of $f(x) = 10x^2$ at the point 2 we simply find $f'(2) = 20(2) = 40$. If we wanted to know the instantaneous rate of change of $f(x)$ at 3 we simply find $f'(3) = 20(3) = 60$. The instantaneous rate of change of $f(x)$ at 1.4 is $f'(1.4) = 20(1.4) = 28$, and so on.

We find the derivative of a function $f(x)$ using the same procedure that was used in the last example. We let $y = f(x)$ and then we let $x_1 = x$ and $x_2 = x + h$. We then find $y_1 = f(x)$ and $y_2 = f(x + h)$ and use the formula for the average rate of change,

$$\frac{y_2 - y_1}{x_2 - x_1} = \frac{f(x + h) - f(x)}{x + h - x} = \frac{f(x + h) - f(x)}{h}.$$

After that, we take the limit as h goes to zero,

$$\lim_{h \to 0} \frac{f(x + h) - f(x)}{h}.$$

The function we obtain after all of this is called the **derivative function of** f or the **derivative of** f or the **derived function of** f. The derivative function of f is often written as f'.

The derivative of f at x is given by

$$f'(x) = \lim_{h \to 0} \frac{f(x+h) - f(x)}{h}.$$

Another way of writing the derivative function is $\frac{df}{dx}$ instead of f'. Calculus was discovered (or invented) independently by different people at roughly the same time; the British mathematician, physicist, and astronomer Sir Isaac Newton and the German mathematician Gottfried Wilhelm Leibniz. These two co-discoverers of calculus invented two different notations for the derivative of a function $f(x)$. Newton used $f'(x)$ and Leibniz used $\frac{df}{dx}$. Both notations are still in use today, so it is important to be able to recognize and use both. To understand the Leibniz notation, consider the formula for the rate of change of y with respect to x,

$$\frac{y_2 - y_1}{x_2 - x_1}.$$

This is the change in y divided by the change in x. The change in y is given by the *difference* in two y values and the change in x is given by the *difference* in two x values. (Recall, differences are found using subtraction.) In science, the Greek capital letter delta, written as Δ, is often used to mean difference. Thus Δy means $y_2 - y_1$ and Δx means $x_2 - x_1$. This allows the formula for the rate of change to be written as

$$\frac{\Delta y}{\Delta x} = \frac{y_2 - y_1}{x_2 - x_1}.$$

When it comes to the derivative formula in the numerator we are finding the change in the function f, so we would write

$$\frac{\Delta f}{\Delta x} = \frac{f(x_2) - f(x_1)}{x_2 - x_1}.$$

If we let $x_1 = x$ and $x_2 = x + h$ and then take the limit at $h \to 0$ then we are taking a very small difference. So, instead of using the Greek capital, or large letter, Δ we simply use a lowercase, or small letter, d,

$$\frac{df}{dx} = \lim_{h \to 0} \frac{f(x+h) - f(x)}{h}.$$

All you really need to remember is that $\frac{df}{dx}$ is simply another way of writing f'. Finally, just so you know, the fraction

$$\frac{f(x+h) - f(x)}{h}$$

is very often simply called the **difference quotient**.

We will now start to develop a number of differentiation rules. It is important to remember that these rules all come from the same place, the limit as h approaches 0 of the difference quotient. In the more straightforward cases, we will provide the proof of the derivative rule, though, in the more complicated situations in later chapters, we will simply provide the rule as some of the proofs become rather involved. Also, in this book, we are mostly interested in using the rules to answer real-world business and economics questions. Once you learn the rule you usually do not return to the definition of the derivative anymore. These rules should all be memorized.

We begin with the simplest rule, the derivative of a constant function. A constant function is a function that is always the same. For example, $f(x) = 5$ is a constant function. This means that no matter what value of x is input into the function the function always outputs the value 5. For example, if $x = 2$ the $f(2) = 5$. If $x = 15$ the $f(15) = 5$. If $x = -7$ then $f(-7) = 5$, and so on.

Example 2.9

Find the derivative of the function $f(x) = 5$.

Solution: We will use our definition of the derivative.

$$
\begin{aligned}
f'(x) &= \lim_{h \to 0} \frac{f(x+h) - f(x)}{h} \\
&= \lim_{h \to 0} \frac{5 - 5}{h} \\
&= \lim_{h \to 0} \frac{0}{h} \\
&= \lim_{h \to 0} (0) \\
&= 0.
\end{aligned}
$$

Notice that zero divided by any number h is still zero. We do this division before substituting 0 into h. By doing this division first the h essentially disappears. Also, even though here we have found that the derivative of the constant function $f(x) = 5$ is $f'(x) = 0$ the same reasoning applies to any constant function.

Derivative of a Constant Function:

If $f(x) = c$, where c is a number, then $f'(x) = 0$.

We now consider some slightly more complicated situations.

Example 2.10

Using the above definition of the derivative, find the derivative of the function $f(x) = x$.

Solution: We note that $f(x + h) = x + h$ and $f(x) = x$, so we have

$$f'(x) = \lim_{h \to 0} \frac{f(x + h) - f(x)}{h} = \lim_{h \to 0} \frac{\not{x} + h - \not{x}}{h}$$

$$= \lim_{h \to 0} \frac{h}{h} = \lim_{h \to 0} \frac{\not{h}}{\not{h}} = \lim_{h \to 0} 1 = 1.$$

Notice that $\lim_{h \to 0} 1 = 1$ since there is no h to substitute 0 into. This is almost silly, but it is important to know that the derivative of $f(x) = x$ is given by the constant function $f'(x) = 1$.

Example 2.11

Using the above definition of the derivative, find the derivative of the function $f(x) = 5x$.

Solution: First note that $f(x + h) = 5(x + h) = 5x + 5h$, so we have

$$f'(x) = \lim_{h \to 0} \frac{f(x + h) - f(x)}{h} = \lim_{h \to 0} \frac{\not{5x} + 5h - \not{5x}}{h}$$

$$= \lim_{h \to 0} \frac{5h}{h} = \lim_{h \to 0} \frac{5\not{h}}{\not{h}} = \lim_{h \to 0} 5 = 5.$$

Thus, we have the derivative of $f(x) = 5x$ is given by $f'(x) = 5$.

Notice, what we did above works for every function of the form $f(x) = mx$, where m is a real number.

Derivative of $f(x) = mx$:

If $f(x) = mx$, where m is a real number, then $f'(x) = m$.

If we combine this with the fact that the derivative of a constant function is zero, we have the following rule:

Derivative of a Linear Function:

If $f(x) = mx + b$, where m and b are real numbers, then $f'(x) = m$.

These two rules are so similar we usually do not think of them as different rules. Now we consider what happens when we take the derivatives of $f(x) = x^2$ and $f(x) = x^3$.

Example 2.12

Using the above definition of the derivative, find the derivative of the function $f(x) = x^2$.

Solution: We first note that $f(x + h) = (x + h)^2 = x^2 + 2xh + h^2$, so we have

$$f'(x) = \lim_{h \to 0} \frac{f(x + h) - f(x)}{h} = \lim_{h \to 0} \frac{x^2 + 2xh + h^2 - x^2}{h}$$

$$= \lim_{h \to 0} \frac{2xh + h^2}{h} = \lim_{h \to 0} \frac{\cancel{h}(2x + h)}{\cancel{h}}$$

$$= \lim_{h \to 0} (2x + h) = 2x.$$

Thus the derivative of $f(x) = x^2$ is $f'(x) = 2x$.

Example 2.13

Using the above definition of derivative find the derivative of the function $f(x) = x^3$.

Solution: We first find

$$f(x + h) = (x + h)^3 = (x + h)(x + h)^2$$
$$= (x + h)(x^2 + 2xh + h^2)$$
$$= x^3 + x^2 h + 2x^2 h + 2xh^2 + xh^2 + h^3$$
$$= x^3 + 3x^2 h + 3xh^2 + h^3.$$

Now we use this in the definition of derivative,

$$f'(x) = \lim_{h \to 0} \frac{f(x + h) - f(x)}{h} = \lim_{h \to 0} \frac{x^3 + 3x^2 h + 3xh^2 + h^3 - x^3}{h}$$

$$= \lim_{h \to 0} \frac{3x^2 h + 3xh^2 + h^3}{h} = \lim_{h \to 0} \frac{\cancel{h}(3x^2 + 3xh + h^2)}{\cancel{h}}$$

$$= \lim_{h \to 0} (3x^2 + 3xh + h^2) = 3x^2.$$

Thus the derivative of $f(x) = x^3$ is $f'(x) = 3x^2$.

Let us make a small table showing what we have found so far,

$f(x)$	$f'(x)$
x	1
x^2	$2x$
x^3	$3x^2$

We can see that a pattern is starting to emerge. The general proof requires the product rule, which is covered later, but this pattern holds in general and gives us the **power rule**. In some sense, the power rule is actually the most important and most used derivative rule in calculus, and it is essential that you know and understand this rule.

The Power Rule:

If $f(x) = x^n$, where n is any real number, then $f'(x) = nx^{n-1}$.

Another way of writing this is $\frac{d}{dx}(x^n) = nx^{n-1}$. This rule not only applies to when n is a positive whole number, but also to when n is a negative number, a fraction, or any real number, though it makes sense to assume that $n \neq 0$.

Even though we reviewed exponents in the introductory chapter, we will go ahead and list the exponential rules here for easy reference. These rules become very useful when taking derivatives. Often we are given expressions with powers of x in the denominator or n^{th} roots. In order to use the power rule we need to write the expression in a "power rule friendly form," which means that we want to write the expression in such a way that we have x raised to a power n. That is where these rules are most helpful.

Exponential Rules: Suppose m and n are real numbers, then

- $x^m \cdot x^n = x^{m+n}$
- $\frac{x^m}{x^n} = x^{m-n}$
- $(x^m)^n = x^{mn}$

- $(xy)^m = x^m y^m$
- $\left(\frac{x}{y}\right)^m = \frac{x^m}{y^m}$

- $x^0 = 1$
- $x^{-m} = \frac{1}{x^m}$
- $x^{\frac{m}{n}} = \sqrt[n]{x^m}$

Example 2.14

Find the derivative of $f(x) = \frac{1}{x^4}$.

Solution: Here we use an exponential rule to write $\frac{1}{x^4} = x^{-4}$, which is our power rule friendly form. We then apply the power rule to get

$$f'(x) = -4x^{-4-1}$$
$$= -4x^{-5}$$
$$= \frac{-4}{x^5}.$$

Notice, we again used an exponential rule to simplify the last step.

Example 2.15

Find the derivative of $f(x) = \sqrt{x}$.

Solution: Here we use an exponential rule to write $\sqrt{x} = x^{\frac{1}{2}}$, which is our power rule friendly form. We then apply the power rule to get $f'(x) = \frac{1}{2}x^{\frac{1}{2}-1} = \frac{1}{2}x^{-\frac{1}{2}} = \frac{1}{2\sqrt{x}}$. Notice, we again used exponential rules to simplify this last step.

Example 2.16

Find the derivative of $f(x) = \sqrt[3]{x^5}$.

Solution: Here we use an exponential rule to write $\sqrt[3]{x^5} = x^{\frac{5}{3}}$, which is our power rule friendly form. We then apply the power rule to get $f'(x) = \frac{5}{3}x^{\frac{5}{3}-1} = \frac{5}{3}x^{\frac{2}{3}} = \frac{5}{3}\sqrt[3]{x^2}$. Again, we used exponential rules to simplify this last step.

Example 2.17

Find the derivative of $f(x) = \frac{1}{\sqrt[3]{x^5}}$.

Solution: Here we use an exponential rule to write $\frac{1}{\sqrt[3]{x^5}} = x^{-\frac{5}{3}}$, which is our power rule friendly form. We then apply the power rule to get $f'(x) = -\frac{5}{3}x^{-\frac{5}{3}-1} = -\frac{5}{3}x^{-\frac{8}{3}} = \frac{-5}{3\sqrt[3]{x^8}}$. Notice both the similarities and differences from the last example.

Finally, we want to emphasize what the derivative is. Recall that when we defined the derivative of a function we used the same procedure that we used when we were finding instantaneous rates of change. Therefore the derivative of a function is a new function that gives the instantaneous rate of change of the original function at each point. The following examples show how this works.

Example 2.18

What is the instantaneous rate of change of $f(x) = x$ at the value $x = 2$? At the value $x = 5$? At the value $x = 10$?

Solution: First we find the derivative of $f(x) = x$ which is $f'(x) = 1$.

- The instantaneous rate of change of $f(x) = x$ at the value $x = 2$ is then given by $f'(2) = 1$.
- The instantaneous rate of change of $f(x) = x$ at the value $x = 5$ is given by $f'(5) = 1$.
- The instantaneous rate of change of $f(x) = x$ at the value $x = 10$ is given by $f'(10) = 1$.

Example 2.19

What is the instantaneous rate of change of $f(x) = x^2$ at the value $x = 2$? At the value $x = 5$? At the value $x = 10$?

Solution: First we find the derivative of the function $f(x) = x^2$ which is given by $f'(x) = 2x$.

- The instantaneous rate of change of $f(x) = x^2$ at the value $x = 2$ is $f'(2) = 2(2) = 4$.
- The instantaneous rate of change of $f(x) = x^2$ at the value $x = 5$ is $f'(5) = 2(5) = 10$.
- The instantaneous rate of change of $f(x) = x^2$ at the value $x = 10$ is $f'(10) = 2(10) = 20$.

2.4 DERIVATIVES OF POLYNOMIALS

We need a few more differentiation rules before differentiation becomes a powerful and useful tool. First of all, suppose we want to know what the derivative of $f(x) = 10x^2$ is. We know from the last section that the derivative of $f(x) = x^2$ is $f'(x) = 2x$, but how do we handle the constant 10 in the front? In essence, we want to know what happens to functions of the form $f(x) = k \cdot g(x)$, where k is a constant.

$$f'(x) = \lim_{h \to 0} \frac{f(x+h) - f(x)}{h}$$

$$= \lim_{h \to 0} \frac{k \cdot g(x+h) - k \cdot g(x)}{h}$$

$$= \lim_{h \to 0} k \cdot \frac{g(x+h) - g(x)}{h}$$

$$= k \cdot \lim_{h \to 0} \frac{g(x+h) - g(x)}{h}$$

$$= k \cdot g'(x)$$

In the fourth line we pulled the constant k out from behind the limit because there was no h we needed to worry about. Thus, we have obtained the next derivative rule.

Derivative of a Constant Times a Function:

Suppose we have a function $f(x) = kg(x)$ defined to be the product of a constant number k, often called a coefficient, and another function $g(x)$. Then the derivative of $f(x)$ is

$$f'(x) = kg'(x).$$

This can also be written as $\frac{d}{dx}(kg) = k\frac{dg}{dx}$. In essence we are "pulling out" the constant coefficient. Now let us look at a concrete example.

Example 2.20

Find the derivative of $f(x) = \frac{1}{5}\sqrt{x}$.

Solution: First we write the function as $f(x) = \frac{1}{5}x^{\frac{1}{2}}$, which is our power rule friendly form. Then we use the above rule, along with the power rule, to get

$$f'(x) = \frac{1}{5} \cdot \frac{1}{2}x^{-\frac{1}{2}} = \frac{1}{10} \cdot \frac{1}{\sqrt{x}} = \frac{1}{10\sqrt{x}}.$$

The next rule states that the derivative of a sum or difference is the sum or difference of the derivatives.

Derivative of the Sum or Difference of Two Functions:

Suppose we have a function $f(x) = g(x) + h(x)$ that is the sum of two functions $g(x)$ and $h(x)$. The derivative of $f(x)$ is

$$f'(x) = g'(x) + h'(x).$$

Suppose we have a function $f(x) = g(x) - h(x)$ that is the difference between two functions $g(x)$ and $h(x)$. The derivative of $f(x)$ is

$$f'(x) = g'(x) - h'(x).$$

This rule is very easy to prove. We will give the proof for the sum of two functions. Suppose that $f(x) = g(x) + h(x)$, where $g(x)$ and $h(x)$ are two functions. We find the derivative of $f(x)$ using the definition of derivative.

$$f'(x) = \lim_{h \to 0} \frac{f(x+h) - f(x)}{h}$$

$$= \lim_{h \to 0} \frac{\left(g(x+h) + h(x+h)\right) - \left(g(x) + h(x)\right)}{h}$$

$$= \lim_{h \to 0} \frac{g(x+h) + h(x+h) - g(x) - h(x)}{h}$$

$$= \lim_{h \to 0} \frac{g(x+h) - g(x) + h(x+h) - h(x)}{h}$$

$$= \lim_{h \to 0} \left(\frac{g(x+h) - g(x)}{h} + \frac{h(x+h) - h(x)}{h}\right)$$

$$= \lim_{h \to 0} \frac{g(x+h) - g(x)}{h} + \lim_{h \to 0} \frac{h(x+h) - h(x)}{h}$$

$$= g'(x) + h'(x)$$

Thus, we have proved that the derivative of the sum of two functions is the sum of the derivatives of each function. Finding the derivative of $f(x) = g(x) - h(x)$ is completely analogous. Also, the rule applies to sums or differences of more than two functions. With this rule, we can now take derivatives of polynomial functions. Basically, we just need to take the derivative of each term in the polynomial separately.

Example 2.21

Find the derivative of $f(x) = 100x - 20x^2$.

Solution: Since the function $f(x)$ is the difference of the functions $100x$ and $20x^2$ we can take the derivative of each of these functions. This amounts to taking the derivative of each term of the polynomial,

$$f'(x) = 100(1) - 20(2x)$$

$$= 100 - 40x.$$

Notice how we use the sum/difference rule, the rule about constant coefficients, and the power rule all at once. It is important to know all these derivative rules and be able to use them simultaneously. From now on we will not point this out.

Example 2.22

Find the derivative of the function

$$f(x) = 10x^7 - 5x^4 + 3x^3 - 6x^2 + 40x - 20.$$

Solution: We simply have to take the derivative of each term in the polynomial, giving us

$$f'(x) = 10(7x^6) - 5(4x^3) + 3(3x^2) - 6(2x) + 40(1) - 0$$
$$= 70x^6 - 20x^3 + 9x^2 - 12x + 40.$$

Example 2.23

Find the derivative of the function $f(x) = 20x^3 - 50x^2 + 100x - 75$ and evaluate it at the value $x = 10$.

Solution: First we find the derivative,

$$f'(x) = 60x^2 - 100x + 100.$$

Next, we evaluate it at $x = 10$. This means finding

$$f'(10) = 60(10)^2 - 100(10) + 100 = 5100.$$

Example 2.24

Find the instantaneous rate of change of $f(x) = 5x^4 - 25x^3 + 100x$ at $x = 5$.

Solution: Recalling that the derivative of a function is a new function that gives the instantaneous rate of change of the original function at each point we first find the derivative of $f(x)$,

$$f'(x) = 20x^3 - 75x^2 + 100.$$

Then to find the instantaneous rate of change of the original function $f(x)$ at $x = 5$ we simply substitute 5 for x in the derived function,

$$f'(5) = 20(5)^3 - 75(5)^2 + 100 = 725.$$

Example 2.25

Find the instantaneous rate of change of $f(x) = \frac{1}{x^3}$ at $x = 2$.

Solution: First we find the derivative of $f(x) = \frac{1}{x^3} = x^{-3}$,

$$f'(x) = -3x^{-4} = \frac{-3}{x^4}$$

and then we substitute $x = 2$ into the derived function,

$$f'(x) = \frac{-3}{x^4} = \frac{-3}{2^4} = \frac{-3}{16} = -0.1875.$$

Taking the derivative of the first derivative gives us the second derivative, taking the derivative of the second derivative gives us the third derivative, and so on. The same rules apply whenever we are looking for the first, second, or higher derivative. We will look into this in detail in Chapter 5.

2.5 GRAPHICAL MEANING OF THE DERIVATIVE

The most important idea we need to understand is the idea of a **tangent line**. Very simply, the tangent line to a function at a point is a line that is the "closest linear approximation" to the function at that point. In other words, it is the line that most closely approximates the function at that point. Graphically, it looks like the tangent line "just touches" the function at that point. Consider the function $f(x) = 5x - 0.1x^2$ shown in the graph below. We have drawn the tangent line to this function at the point $x = 10$.

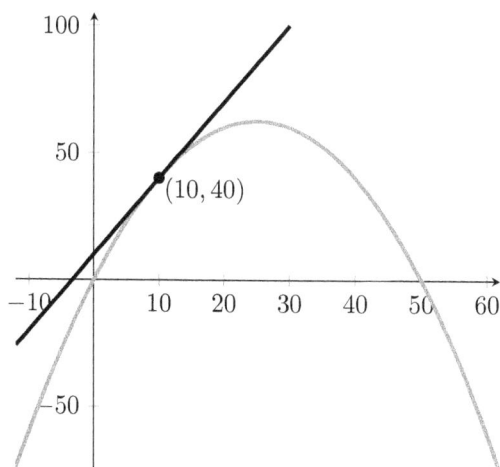

Notice the black dot. It is on the graph of $f(x) = 5x - 0.1x^2$. In other words, it has an x value of 10 and a y value of $f(10) = 40$. Therefore it would be more accurate to say that the black tangent line "just touches" the graph at the point $(10, 40)$. Also notice that for x values close to 10 the y values on the tangent line are very close approximations to the y values on the function. Sometimes we will say the black line is the tangent line to $f(x)$ at the point $x = 10$ and sometimes we will say it is the tangent line to $f(x)$ at the point $(10, 40)$. These are simply two ways of saying the same thing.

Of course, the graph of a function has a tangent line at every point of the function. Below we show five different tangent lines to the graph $f(x) = 5x - 0.1x^2$ at the five points $x = 10$, $x = 20$, $x = 25$, $x = 35$ and $x = 45$.

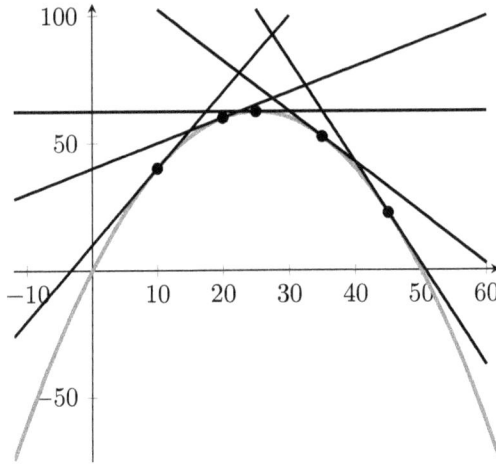

The question now is, how exactly do we get a tangent line? Above we said that a tangent line to a function at a point is a line that "just touches" the graph of the function at that point. But this is not precise. We need a better idea of how tangent lines are found. It turns out that tangent lines are closely related to instantaneous rates of change.

Recall in section 2.1 we learned about average rates of change. Suppose we wanted to know the average rate of change of the function $f(x) = 5x - 0.1x^2$ with respect to x as x changes from 10 to 30. We would first find $f(10) = 40$ and $f(30) = 60$ and then calculate

$$\frac{y_2 - y_1}{x_2 - x_1} = \frac{f(30) - f(10)}{30 - 10} = \frac{60 - 40}{30 - 10} = \frac{20}{20} = 1.$$

This formula is also the formula for finding the slope of the line segment that connects the two points

$$(x_1, y_1) = (x_1, f(x_1)) = (10, f(10)) = (10, 40),$$
$$(x_2, y_2) = (x_2, f(x_2)) = (30, f(30)) = (30, 60).$$

Below we show the line segment that connects the two points $(10, 40)$ and $(30, 60)$ in gray. The line that connects these two points is shown in black.

Thus, the average rate of change between $x = 10$ and $x = 30$ is actually the slope of this line.

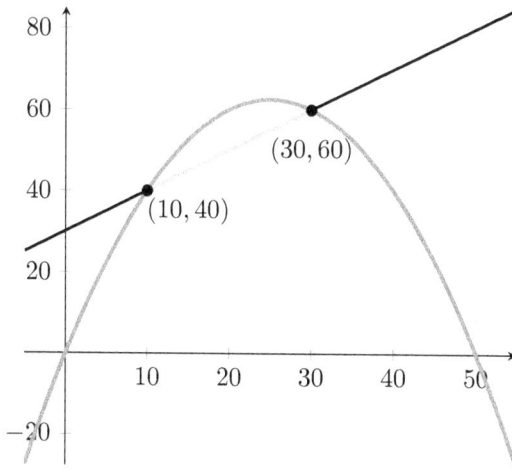

After learning about average rates of change we then learned about instantaneous rates of change in section 2.2. If we wanted to know the instantaneous rate of change of $f(x)$ at $x_1 = 10$ we let x_2 get closer and closer to $x_1 = 10$. For example, if we chose $x_2 = 20$, then we have

$$\frac{y_2 - y_1}{x_2 - x_1} = \frac{f(20) - f(10)}{20 - 10} = \frac{60 - 40}{20 - 10} = \frac{20}{10} = 2.$$

This gives us the slope of the line segment that connects the points

$$(x_1, y_1) = (x_1, f(x_1)) = (10, f(10)) = (10, 40),$$
$$(x_2, y_2) = (x_2, f(x_2)) = (20, f(20)) = (20, 60).$$

This line segment is shown in gray and the line is shown in black below.

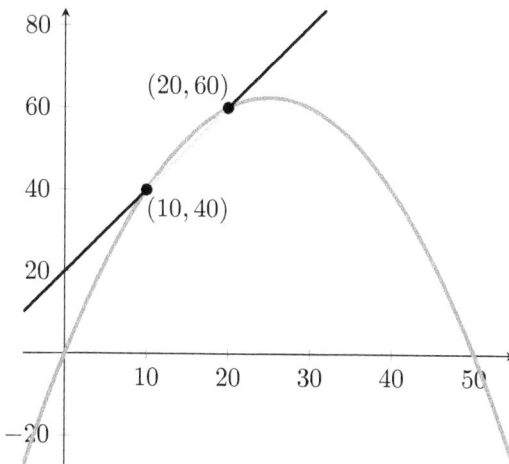

We could choose an x_2 value even closer to $x_1 = 10$. Suppose we chose $x_2 = 12$. Then we would find

$$\frac{y_2 - y_1}{x_2 - x_1} = \frac{f(12) - f(10)}{12 - 10} = \frac{45.6 - 40}{12 - 10} = \frac{5.6}{2} = 2.8$$

which is the slope of the line segment that connects the points $(10, 40)$ and $(12, 45.6)$. This line segment is shown below in gray and the full line is shown in black.

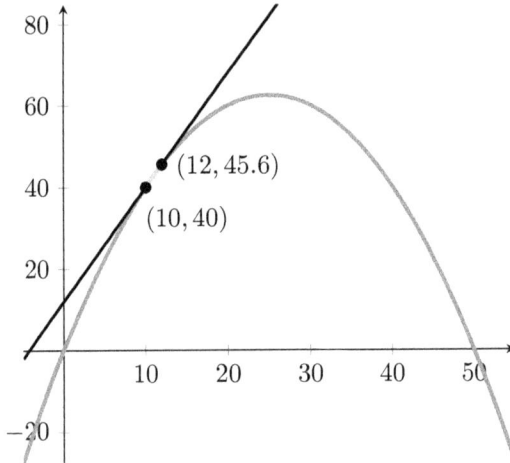

In section 2.2 we learned about instantaneous rates of change. We found the rate of change

$$\frac{y_2 - y_1}{x_2 - x_1} = \frac{f(x_2) - f(x_1)}{x_2 - x_1}$$

as x_2 got closer and closer to x_1. The easiest way to do this was to let $x_2 = x_1 + h$ and then take the limit as h approaches zero. In other words, the instantaneous rate of change of the function $f(x)$ at the point x_1 is given by

$$\lim_{h \to 0} \frac{f(x_1 + h) - f(x_1)}{(x_1 + h) - x_1}.$$

This simplifies to

$$\lim_{h \to 0} \frac{f(x_1 + h) - f(x_1)}{h}.$$

For each value of h, we have a small line segment that connects the points at $x_2 = x_1 + h$ and x_1. Also, each of these line segments determines a line. The line we obtain when h goes to zero is called the tangent line. Below we show the graph of $f(x) = 5x - 0.1x^2$ in gray, along with the several lines we have found above, also in gray. As $h \to 0$ then these lines approach the tangent line at $x = 10$, which is shown in black.

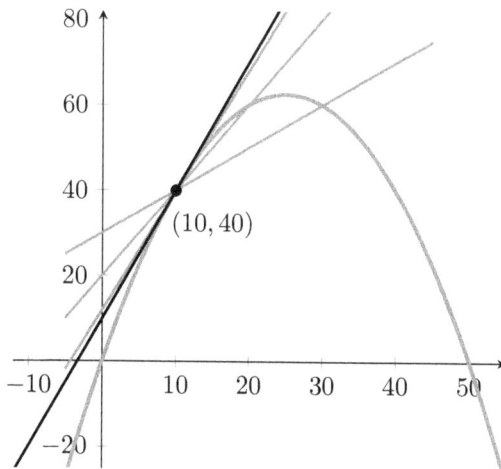

In 2.3 we learned that

$$\lim_{h \to 0} \frac{f(x_1 + h) - f(x_1)}{h}.$$

was called the derivative of the function f at the point x_1. This means that **the slope of the tangent line at the point x_1 is given by the derivative of $f(x)$ at the point x_1**. This is an incredibly useful fact, as we will soon see.

2.6 APPLICATIONS OF RATES OF CHANGE AND DERIVATIVES IN BUSINESS

Since this is a course on calculus and its applications in both business and economics, all of our real-world examples will come from either of these fields. Therefore, it is not only important to understand the mathematics necessary to solve the problem, but it is also important to understand and interpret, in terms of business concepts, what the answer means.

Example 2.26

Suppose a profit function for the production of widgets is given by $P(x) = -2x^2 + 400x - 500$. If the production of widgets increases from $x = 40$ to $x = 50$, what is the average rate of change in profit?

Solution: We first let

$$x_1 = 40 \text{ widgets,}$$
$$x_2 = 50 \text{ widgets.}$$

Next, we have to find the values of y,

$$y_1 = P(x_1) = P(40) = -2(40)^2 + 400(40) - 500 = 12300,$$
$$y_2 = P(x_2) = P(50) = -2(50)^2 + 400(50) - 500 = 14500.$$

The average rate of change in profit when production increases from 40 to 50 is

$$\frac{y_2 - y_1}{x_2 - x_1} = \frac{14500 - 12300}{50 - 40} = 220.$$

But what are the units of this answer? We can assume that the profit is given in terms of dollars and we are producing widgets, thus the units of the answer is actually 220 dollars per widget, which can be written as

$$220 \frac{\text{dollars}}{\text{widget}}.$$

We note, widgets is simply a generic term used to mean some unnamed device or gadget. Many of our examples involve widgets. Also, keeping track of the correct units is very important in business, therefore when we are dealing with business examples we will try to do that. However, when we are focusing on mathematical topics we often will not pay such close attention to the units. Hopefully, this will not create too much confusion.

Example 2.27

The revenue for a company that makes widgets is given by the revenue function $R(x) = 500x - 10x^2$ where x is the number of widgets produced and the revenue is in thousands of dollars. What is the average rate of change in revenue with respect to widgets produced as production increases from 10 widgets to 20 widgets.

Solution: Here x is the number of widgets produced and $y = R(x)$ is the revenue in thousands of dollars. We are also told production increases from 10 widgets to 20 widgets, so 10 is the initial, or first, value of x, and 20 is the final, or second, value of x,

$$x_1 = 10 \text{ widgets,}$$
$$x_2 = 20 \text{ widgets.}$$

We use these values of x to find the associated values of y,

$$y_1 = R(x_1) = R(10) = 500(10) - 10(10)^2 = 4000 \text{ thousand dollars,}$$
$$y_2 = R(x_2) = R(20) = 500(20) - 10(20)^2 = 6000 \text{ thousand dollars.}$$

Thus the average rate of change in revenue with respect to widgets produced as production increases from 10 widgets to 20 widgets is

$$\frac{y_2 - y_1}{x_2 - x_1} = \frac{6000 \text{ thousand dollars} - 4000 \text{ thousand dollars}}{20 \text{ widgets} - 10 \text{ widget}}$$

$$= 200 \frac{\text{thousand dollars}}{\text{widget}}.$$

The average rate of change in revenue with respect to widgets as production increases from 10 widgets to 20 widgets is 200,000 dollars per widget.

Example 2.28

Using what you found in the last example, estimate how much would you expect revenue to change if your production increased from 11 widgets to 12 widgets? If production increased from 16 widgets to 18 widgets? If production increased from 19 widgets to 21 widgets?

Solution: In the last example we found that the average rate of change in revenue with respect to widgets as production increases from 10 widgets to 20 widgets is 200,000 dollars per widget.

- Since both 11 and 12 are within the range from 10 to 20 the average range of change we found applies as we increase production from 11 widgets to 12 widgets. Our increase is $12 - 11 = 1$ widget so our estimated increase in revenue is given by

$$200{,}000 \frac{\text{dollars}}{\text{widget}} \cdot 1 \text{ widget} = 200{,}000 \text{ dollars}.$$

- Since both 16 and 18 are within the range from 10 to 20 the average range of change we found applies as we increase production from 16 widgets to 18 widgets. Our increase is $18 - 16 = 2$ widgets so our estimated increase in revenue is given by

$$200{,}000 \frac{\text{dollars}}{\text{widget}} \cdot 2 \text{ widgets} = 400{,}000 \text{ dollars}.$$

- Since 21 is not within the range from 10 to 20 so the average rate of change we found in the last example does not apply as we increase production from 19 widgets to 21 widgets thus we can not answer this question.

In the last example, we have used the average rate of change of revenue with respect to widgets produced, as widget production increases from 10 to 20 widgets, to **estimate** the change in revenue we would expect if we increased production from 11 to 12 widgets. However, this is only an estimate. Suppose we wanted to know the **exact** change in revenue.

Example 2.29

Using the revenue function $R(x) = 500x - 10x^2$, where x is the number of widgets produced and the revenue is in thousands of dollars, find the exact change in revenue if production increased from 11 widgets to 12 widgets. If production increased from 16 widgets to 18 widgets? If production increased from 19 widgets to 21 widgets?

Solution:

- To find the exact change in revenue as production increases from 11 to 12 widgets we first need to find the revenue when production is 11 widgets and the revenue when production is 12 widgets.

$$R(11) = 500(11) - 10(11)^2 = 4290 \text{ thousand dollars},$$
$$R(12) = 500(12) - 10(12)^2 = 4560 \text{ thousand dollars}$$

so the exact change in revenue as production increases from 11 to 12 widgets is given by

$$R(12) - R(11) = 4560 \text{ thousand dollars} - 4290 \text{ thousand dollars}$$
$$= 270 \text{ thousand dollars}.$$

- To find the exact change in revenue as production increases from 16 to 18 widgets we first need to find the revenue when production is 16 widgets and the revenue when production is 18 widgets.

$$R(16) = 500(16) - 10(16)^2 = 5440 \text{ thousand dollars},$$
$$R(18) = 500(18) - 10(18)^2 = 5760 \text{ thousand dollars}$$

so the exact change in revenue as production increases from 16 to 18 widgets is given by

$$R(18) - R(16) = 5760 \text{ thousand dollars} - 5440 \text{ thousand dollars}$$
$$= 320 \text{ thousand dollars}.$$

- To find the exact change in revenue as production increases from 19 to 21 widgets we first need to find the revenue when production is 19 widgets and the revenue when production is 21 widgets.

$$R(19) = 500(19) - 10(19)^2 = 5890 \text{ thousand dollars,}$$
$$R(21) = 500(21) - 10(21)^2 = 6090 \text{ thousand dollars}$$

so the exact change in revenue as production increases from 16 to 18 widgets is given by

$$R(21) - R(19) = 6090 \text{ thousand dollars} - 5890 \text{ thousand dollars}$$
$$= 200 \text{ thousand dollars.}$$

Notice the answers we get when we found the exact change in revenue are different than the answers we got when we estimated the change in revenue by using the average rate of change. This is to be expected.

In business, the word **marginal** is used instead of the phrase *instantaneous rate of change*. Thus, if one is given a cost function and is asked what the marginal cost is at a some value this is the same thing as asking what the instantaneous rate of change is at that value. We simply have to find the derivative of the cost function and substitute the value into the derived function. Marginal revenue and marginal profit are of course defined exactly the same way.

Example 2.30

Suppose the cost in dollars of making a certain product is given by the cost function $C(x) = 350x - 0.75x^2$. What is the marginal cost when production is 150?

Solution: The first thing we do is take the derivative,

$$C'(x) = 350 - 1.5x.$$

This is called the marginal cost function. To find the marginal cost when production is 150 we substitute the number 150 into the derived function,

$$C'(150) = 350 - 1.5(150) = 125.$$

So the marginal revenue of making this product when production is 150 is 125 dollars. In other words, the instantaneous rate of change in cost when production is 150 is 125 dollars.

Example 2.31

The revenue in thousands of dollars obtained from producing x units of some item is given by $R(x) = 5x - 0.0005x^2$. Find the marginal revenue at a production level of $x = 1000$ items.

Solution: First we find the derivative of the revenue function,

$$R'(x) = 5 - 0.001x.$$

When the production level is $x = 1000$, then we have

$$R'(1000) = 5 - 0.001(1000) = 4.$$

But remember, the revenue is given in thousands of dollars. This means that the marginal revenue at $x = 1000$ is 4 thousand dollars, or 4000 dollars. Another way of saying this is that the instantaneous rate of change in revenue when production is 1000 is 4000 dollars.

Example 2.32

Suppose that the total profit in dollars from selling x items is given by $P(x) = 4x^2 - 5x + 10$. Find the marginal profit at $x = 5$.

Solution: First take the derivative,

$$P'(x) = 8x - 5.$$

Evaluating the derivative at $x = 5$ gives us

$$P'(5) = 8(5) - 5 = 35.$$

So the marginal profit when sales are $x = 5$ is 35 dollars.

But why is marginal cost, revenue, or profit important? What does it mean? How do business people and economists use it? Suppose we had the following revenue function:

$$R(x) = 200x - 0.1x^2.$$

What is the increase in revenue when production increases from 300 units to 301 units? First we find the revenue when production is both 301 and 300 units,

$$R(301) = 200(301) - 0.1(301)^2 = 51139.90,$$
$$R(300) = 200(300) - 0.1(300)^2 = 51000.$$

The increase in revenue as production increases from 300 units to 301 units is given by

$$R(301) - R(300) = 51139.90 - 51000 = 139.90.$$

Now we will find the marginal revenue when production is 300. For this, we first find the marginal revenue function, which is nothing more than the derivative of the revenue function,

$$R'(x) = 200 - 0.2x.$$

Then we substitute in the value 300,

$$R'(300) = 200 - 0.2(300) = 140.$$

The answer we obtained when we found the marginal revenue at 300 was approximately equal to the change in revenue when production increased from 300 to 301. This is why marginal revenue is important. **Marginal revenue at x is the approximate expected increase in revenue when sales go from x items to $x+1$ items.** Similarly, marginal cost at x is the approximate expected increase in cost when production goes from x items to $x + 1$ items, and marginal profit at x is the approximate expected increase in profit when sales go from x items to $x + 1$ items.

Example 2.33

The total cost in dollars to produce x items is given by the cost function $C(x) = 60 + 2x - x^2 + 5x^3$. Find the marginal cost when production is $x = 5$ and explain what this means.

Solution: First we take the derivative of the cost function,

$$C'(x) = 2 - 2x + 15x^2.$$

To find the marginal cost at a production level of $x = 5$ we substitute 5 into the derivative formula,

$$C'(5) = 2 - 2(5) + 15(5)^2 = 367.$$

The marginal cost when $x = 5$ is 367 dollars. This means that the instantaneous rate of change in cost at $x = 5$ is given by 367 dollars. The instantaneous rate of change at $x = 5$ can be used to approximate the rate of change as x changes from 5 to 6. This means that the change in cost as production rises from 5 to 6 is approximately 367 dollars.

Example 2.34

Suppose the total cost of producing x widgets is given by the cost function $C(x) = 100 + 150x - 0.09x^2$. What is the exact cost of producing the 65th widget? Use the marginal cost function to approximate the cost of producing the 65th widget.

Solution: The cost of producing the 65th widget is the cost increase in cost from producing 64 widgets to 65 widgets. Finding

$$C(64) = 100 + 150(64) - 0.09(64)^2 = 9331.36,$$
$$C(65) = 100 + 150(65) - 0.09(65)^2 = 9469.75.$$

Thus the increase in cost is given by

$$C(65) - C(64) = 9469.75 - 9331.36 = 138.39.$$

Thus the exact cost of producing the 65th widget is 138.39. We can use the marginal cost function to approximate the exact cost. We first find the marginal cost function by taking the derivative of the cost function,

$$C'(x) = 150 - 0.18x,$$

which we then evaluate at 64,

$$C'(64) = 150 - 0.18(64) = 138.48.$$

The marginal cost when production is 64 is 138.48, which is an approximation of the cost of producing the 65th widget.

A quick word on notation, economists and business people often use the initials MC to refer to marginal costs, MR to refer to marginal revenue, MP to refer to marginal profits, and so on. Similarly, they also use TC for total costs, or what a mathematician would usually write as the cost function $C(x)$, TR for total revenues, or $R(x)$, and TP for total profits, or $P(x)$. Similarly, FC refers to fixed costs and VC refers to variable costs. It is important to become comfortable with this notation as well as the more standard mathematical notations.

Economists are often interested in knowing the average value of many economic variables. For example, an economist may want to know the **average revenue** generated by selling x items. Therefore, the average revenue, denoted as AR, is defined as the average revenue per unit sold for the first x units sold. This is obtained by taking the total revenue TR obtained from selling x items and dividing that number by x,

$$AR = \frac{TR}{x}.$$

A mathematician would usually write this as

$$\overline{R}(x) = \frac{R(x)}{x},$$

where the bar above the R is the mathematical notation that means average. If we consider this last equation and cross multiply, we find the following:

$$AR = \frac{TR}{x} \implies TR = AR * x,$$

as we know from before $TR = p * x$, therefore,

$$AR * \not{x} = p * \not{x}. \implies AR = p$$

So, the AR is equal to price p, where p is given by the price-demand function $p = f(x)$. Notice that here we used the symbol $*$ to represent multiplication. This is the symbol generally used for multiplication in economics. We also avoid using the $\times$ symbol since it is so similar to the variable x.

Similar to average revenue, we can define **average cost**, denoted AC, which is defined as the total cost divided by the level of output,

$$AC = \frac{TC}{x} \quad \text{or} \quad \overline{C}(x) = \frac{C(x)}{x}.$$

Since $TC = FC + VC$, we can write the above equation as:

$$AC = \frac{TC}{x} = \frac{FC + VC}{x} = \frac{FC}{x} + \frac{VC}{x} = AFC + AVC$$

where AFC represent the average fixed cost defined as $\frac{FC}{x}$ and AVC represents the average variable cost defined as $\frac{VC}{x}$.

Example 2.35

The demand function of a firm is given by $p = 10 - x$. Find TR and AR.

Solution: We first find total revenue,

$$TR = p * x \implies TR = (10 - x) * x \implies TR = 10x - x^2.$$

This can also be written as $R(x) = 10x - x^2$. Next, we find the average revenue,

$$AR = \frac{TR}{x} = \frac{10x - x^2}{x} = \frac{\not{x}(10 - x)}{\not{x}} = 10 - x$$

which can also be written as $\overline{R}(x) = 10 - x$. Notice that $AR = 10 - x = p$; as noted earlier, AR is the p defined by the price-demand function.

Example 2.36

Find TC and MC given $AC = 3x^2 - 4x + 2 + \frac{10}{x}$.

Solution:

$$AC = \frac{TC}{x} \implies TC = AC * x$$

$$\implies TC = \left(3x^2 - 4x + 2 + \frac{10}{x}\right) * x$$

$$\implies TC = 3x^3 - 4x^2 + 2x + 10$$

Then,

$$MC = TC' \implies MC = \left(3x^3 - 4x^2 + 2x + 10\right)'$$

$$\implies MC = 9x^2 - 8x + 2$$

Example 2.37

Consider the following average cost function $AC = x^2 - 5x + \frac{15}{x} + 60$.

1. Find TC and evaluate it when quantity demanded is 8.
2. Find FC and VC.
3. Find MC.

Solution:

1. To find TC we multiply AC by x as follows:

$$TC = AC * x = \left(x^2 - 5x + \frac{15}{x} + 60\right) * x$$

$$= x^3 - 5x^2 + 15 + 60x$$

Substituting $x = 8$ will give us TC when quantity demanded is

$$TC = 8^3 - 5(8)^2 + 15 + 60(8) = 687.$$

2. As $TC = x^3 - 5x^2 + 15 + 60x$ by rearranging the terms we obtain $TC = 15 + (x^3 - 5x^2 + 60x)$. Taking into consideration that FC does not depend on the input, while VC does, we obtain $FC = 15$ and $VC = x^3 - 5x^2 + 60x$.
3. As $MC = TC'$, we obtain

$$MC = \left(x^3 - 5x^2 + 15 + 60x\right)' = 3x^2 - 10x + 60.$$

We would like to conclude this chapter by providing a summary of the four different ways in which we can talk about, or describe, derivatives. In other words, we have four different sets of language which all describe the exact same concept. It is important to recognize that these are all different ways of saying exactly the same thing.

Four different ways to describe the derivative of a function $f(x)$ at $x = a$

1. The derivative of $f(x)$ at the point $x = a$, written as $f'(a)$.
2. The instantaneous rate of change of $f(x)$ at $x = a$.
3. The slope of the tangent line to $f(x)$ at $x = a$.
4. The marginal cost/revenue/profit of $C(x)/R(x)/P(x)$ at $x = a$.

Since marginal is an economics concept, it is usually only used in relation to cost, revenue, or profit functions.

2.7 PROBLEMS

Question 2.1 *Let $y = 10x^2$. Find the average rate of change of y with respect to x as*

(a) *x changes from 1 to 3,*

(b) *x changes from 5 to 8,*

(c) *x changes from 4 to 6.*

Question 2.2 *Let $y = f(x)$ where $f(x) = 100x - 20x^2$. Find the average rate of change of y with respect to x as*

(a) *x changes from 0 to 2,*

(b) *x changes from 2 to 3,*

(c) *x changes from 3 to 5.*

Question 2.3 *Find the average rate of change of the function $f(x) = 5x^2 + 8x$ over the interval $x = -4$ to $x = 3$.*

Question 2.4 *Find the average rate of change of the function $f(x) = 4x^3 + 7x^2 + 5$ over the interval between $x = -3$ and $x = 4$.*

Question 2.5 *Let $y = 2x^2$. Find the instantaneous rate of change of y with respect to x at a value x (here we let the variable x stand for a number). Do this by letting the first value of x be $x_1 = x$ and the second value of x be $x_2 = x + h$. Use the formula for the average rate of change and then take the limit as h goes to zero.*

Question 2.6 Let $y = -50x^2$. Find the instantaneous rate of change of y with respect to x at a value x (here we let the variable x stand for a number). Do this by letting the first value of x be $x_1 = x$ and the second value of x be $x_2 = x + h$. Use the formula for the average rate of change and then take the limit as h goes to zero.

Question 2.7 Find the instantaneous rate of change of $f(x) = x^3$ at

(a) $x = 3$,   (b) $x = 5$,   (c) $x = 7$.

Question 2.8 Find the instantaneous rate of change of $f(x) = x^7$ at

(a) $x = 1$,   (b) $x = 2$,   (c) $x = 3$.

Question 2.9 Find the instantaneous rate of change of $f(x) = x$ at

(a) $x = 2$,   (b) $x = 4$,   (c) $x = 6$.

Question 2.10 Find the instantaneous rate of change of $f(x) = 3x^2$ at

(a) $x = 1$,   (b) $x = 5$,   (c) $x = 6$.

Question 2.11 Find the instantaneous rate of change of $f(x) = 5x^4$ at

(a) $x = 0$,   (b) $x = 2$,   (c) $x = 3$.

Question 2.12 Find the instantaneous rate of change of $f(x) = -2x^3$ at

(a) $x = 1$,   (b) $x = 3$,   (c) $x = 7$.

Question 2.13 Find the instantaneous rate of change of $f(x) = \frac{1}{x^2} = x^{-2}$ at

(a) $x = 1$,   (b) $x = 2$,   (c) $x = 5$.

Question 2.14 Find the instantaneous rate of change of $f(x) = \frac{1}{x^5} = x^{-5}$ at

(a) $x = 1$,   (b) $x = 3$,   (c) $x = 5$.

Question 2.15 Find the instantaneous rate of change of $f(x) = \frac{5}{x^3} = 5x^{-3}$ at

(a) $x = 2$,   (b) $x = 4$,   (c) $x = 5$.

Question 2.16 *Find* $\lim_{x \to -4} \frac{4-x^2}{x+4}$.

Question 2.17 *Find* $\lim_{x \to 0} \frac{x^3 - 3x}{2x}$.

Question 2.18 *Find* $\lim_{x \to 2} \frac{3x^2 - 5x + 6}{x - 2}$.

Question 2.19 *Find the derivative of the function* $f(x) = x^3$ *and evaluate it at* $x = 6$.

Question 2.20 *If* $f(x) = \sqrt{x} = x^{\frac{1}{2}}$ *find* $f'(25)$.

Question 2.21 *If* $f(x) = 14x^5 + 7x^3 - 10x$ *find* $f'(x)$.

Question 2.22 *If* $f(x) = 9x^{-3} + 12x^{-2} - 10x^{-1}$ *find* $f'(x)$.

Question 2.23 *Evaluate the derivative of* $f(x) = 15x^3 - 5x^2$ *at* $x = 7$.

Question 2.24 *If* $f(x) = -4x^5 + 8x^3 - 7x^2$ *find* $f'(5)$.

Question 2.25 *Find the instantaneous rate of change for the function* $f(x) = 6x^2 + 2x$ *at* $x = 3$.

Question 2.26 *Find the instantaneous rate of change for the function* $f(x) = 8x^3 - 4x^2 + 5$ *at* $x = 7$.

Question 2.27 *Simplify then find the derivative of* $f(x) = \frac{x^{10}}{x^2}$.

Question 2.28 *Simplify then find the derivative of* $f(x) = \left(\frac{x^2 + 3x}{4x}\right) + 5$.

Question 2.29 *Simplify then find the derivative of* $\frac{\sqrt{x}}{x}$.

Question 2.30 *Find the derivative of* $\frac{\sqrt{x}}{2}$.

Question 2.31 *Find the derivative of* q^3.

Question 2.32 *Find the derivative of the following total cost function:* $TC = \frac{x^3}{4} - 5x^2 + 2x + 100$.

Question 2.33 *Find the derivative of the following total revenue function:* $TR = (340 - 20x)x$.

Question 2.34 *The total number (in hundreds) of visitors to the aquarium is given by:*

$$N(t) = 3t + \frac{1}{\sqrt{t^3}}$$

where t *is the number of years since 2000. Find* $N(22)$ *and* $N'(22)$ *and interpret both numbers.*

Question 2.35 *The resale price of used textbooks decreases with time* t *in years. Its price* P *is given as*

$$P(t) = 150 - 23\sqrt{\frac{t}{2}}.$$

Find the rate of decrease in the price after 5 years.

Question 2.36 *The demand function for a commodity is* $p = 90 - q^{\frac{5}{2}}$.

1. *Find the total revenue, the marginal revenue, and the average revenue equations.*
2. *What is the slope of the tangent line of the marginal revenue compared to the tangent line of the average revenue?*
3. *find the quantity for which the marginal revenue is equal to zero?*
4. *find the quantity for which the average revenue is equal to zero?*

Question 2.37 *The average cost of a firm is giving as* $AC = q^2 - 7q + \frac{120}{q} + 60$.

1. *Find the equations of the* TC.
2. *What is the* TC *when* $q = 15$?
3. *What is the* FC?
4. *What is the* VC?
5. *Find the* MC.

CHAPTER 3

Exponential and Logarithmic Derivatives

In this chapter, we will look at other types of functions that we did not explore in the first two chapters. We will be introducing exponential and logarithmic functions, their properties, derivatives, and applications in business and economics.

3.1 EXPONENTIAL FUNCTIONS

Exponential functions are functions that have the variable x in the exponent. Exponential functions have the form

$$f(x) = b^x$$

where $b > 0$. The constant b is often called the **base** of the exponential function. Notice, the base must be positive. Exponential functions could also have a constant in front, so sometimes you may see them in the form

$$f(x) = c \cdot b^x$$

where c is a constant and $b > 0$.

Example 3.1

Some examples of exponential functions.

- $f(x) = 2^x$
- $f(x) = 3^x$
- $f(x) = 4 \cdot 2^x$
- $f(x) = -7 \cdot 3^x$

- $f(x) = \left(\frac{1}{2}\right)^x$
- $f(x) = \left(\frac{1}{3}\right)^x$
- $f(x) = 2 \cdot \left(\frac{1}{2}\right)^x$
- $f(x) = \frac{-2}{3} \cdot \left(\frac{1}{3}\right)^x$

Example 3.2

Graph the exponential functions $f(x) = 2^x$, $f(x) = 3^x$, and $f(x) = 4^x$.

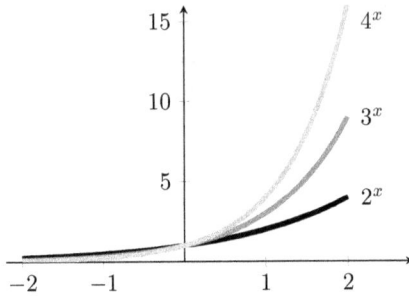

Look carefully at the graphs of the exponential functions. The graphs of exponential functions $f(x) = b^x$, where $b > 1$, all look very similar. They all have the following properties:

1. The function is always positive.
2. The function is always increasing.
3. When $x = 0$ then $f(x) = 1$. (Since any number raised to the zeroth power is one.)
4. The derivative at any point is positive.
5. The slope of the tangent line to the graph of the function at any given point is positive.

But above when we defined the exponential functions we said that $b > 0$. What happens with the exponential function $f(x) = 1^x$?

Example 3.3

Graph the exponential function $f(x) = 1^x$.

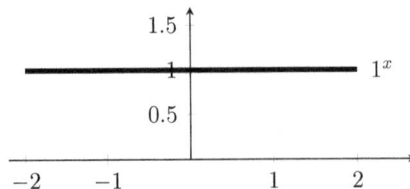

The graph of the exponential function $f(x) = 1^x$ is simply a horizontal line at $y = 1$.

And now what happens when $0 < b < 1$?

Example 3.4

Graph the exponential functions $f(x) = \left(\frac{1}{2}\right)^x$, $f(x) = \left(\frac{1}{3}\right)^x$, and $f(x) = \left(\frac{1}{4}\right)^x$.

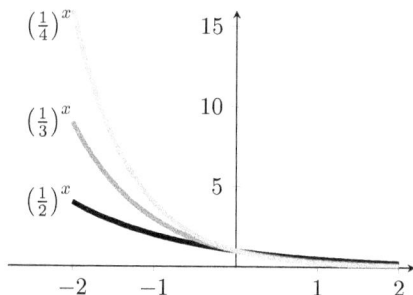

The graphs of exponential functions $f(x) = b^x$, where $0 < b < 1$, all look very similar too. They all have the following properties:

1. The function is always positive.
2. The function is always decreasing.
3. When $x = 0$ then $f(x) = 1$. (Since any number raised to the zeroth power is one.)
4. The derivative at any point is negative.
5. The slope of the tangent line to the graph of the function at any given point is negative.

Example 3.5

Compare the graphs of the exponential functions $f(x) = 3^x$ and $f(x) = \left(\frac{1}{3}\right)^x$.

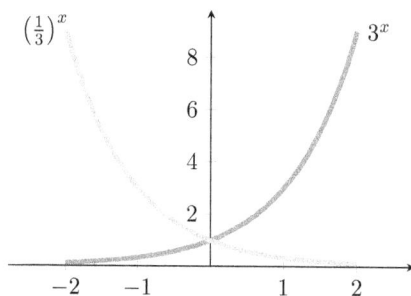

Notice these two graphs are the reflections of each other across the y-axis.

There is one exponential function that is extremely important. It is the exponential function

$$f(x) = e^x.$$

Here e is a number. The number e is a lot like the number π. The number e can actually be defined in several different ways. One way it is defined is

$$e = \lim_{n \to \infty} \left(1 + \frac{1}{n}\right)^n.$$

When e is written in decimal form the decimals go on and on forever and never repeat. For example, the first 225 digits of e are given by

$e = 2.718281828459045235360287471352662497757247093699959957496$

69676277240766303535475945713821785251664274274663919320

03059921817413596629043572900334295260595630738132328627

9434907632338298807531952510190115738341879307021540 89149...

but usually, we are perfectly happy with the first nine digits given by our calculators. In any event, the number e is important for the following fact:

The derivative of the exponential function $f(x) = e^x$ is given by

$$f'(x) = e^x.$$

Yes, the derivative of the exponential function $f(x) = e^x$ is exactly the same as the original function. This is one of the reasons that the number e is so important. We will not prove this fact. Before giving the rule for the derivatives of the exponential functions $f(x) = b^x$, where $b \neq e$ we first need to introduce logarithmic functions.

3.2 LOGARITHMIC FUNCTIONS

In a sense **logarithmic functions** are the inverse of exponential functions. If we asked what 2^4 was it would be easy to calculate $2^4 = 2 \cdot 2 \cdot 2 \cdot 2 \cdot = 16$. But suppose instead we asked what power does 2 need to be raised to to get 16. In other words, find y when

$$2^y = 16.$$

The y is given by a logarithmic function. Here we would have

$$y = \log_2(16) = 4.$$

Notice that there is a little subscript 2 after the log. This 2 tells us the **base** of the logarithm function. The base is the number that is getting raised to the power y.

Logarithmic functions: If $b^y = x$ with $b > 0$ and $b \neq 1$ then

$$y = \log_b(x).$$

For each base b there is a different logarithm function $\log_b$. Usually $\log_{10}$ is simply written as log.

Example 3.6

Examples of Logarithm functions.

- $f(x) = \log_2(x)$
- $f(x) = \log_3(x)$

- $f(x) = \log_4(x)$
- $f(x) = \log_{10}(x) = \log(x)$

Before continuing, we will briefly provide the logarithm rules. These rules are very useful in the algebraic manipulations that are necessary in solving a variety of problems. We will not attempt to prove these rules here, but they are essentially the logarithmic equivalent of the exponential rules.

Logarithm Rules: Suppose n is a real number, then

- $\log_b(1) = 0$
- $\log_b(b) = 1$
- $\log_b(b^n) = n$

- $\log_b(x^n) = n \cdot \log_b(x)$
- $\log_b(x \cdot y) = \log_b(x) + \log_b(y)$
- $\log_b\left(\frac{x}{y}\right) = \log_b(x) - \log_b(y)$

It is important to recognize that b can be any base, and in particular one of the most common bases is e. If $b = e$ then we write $\log_e$ as ln. This abbreviation comes from the Latin logarithmus naturali, which in English is called the **natural logarithm function**. This logarithm is called "natural" because of the wide range of real-world applications the base e appears in.

Natural logarithm function: If $e^y = x$ then

$$y = \ln(x).$$

In this case the rule $\log_b(b^n) = n$ would be written as $\ln(e^n) = n$. This is probably the most commonly used rule in simplifying algebraic expressions that involve natural logarithms. The second most commonly used rule is probably $\log_b(x^n) = n \log_b(x)$, which is written $\ln(x^n) = n \ln(x)$ for natural logarithms.

Example 3.7

Solve $7 = 4^x$ for x.

Solution: This is a very common type of problem one encounters, and the following strategy is almost always used. First, we take the natural log of both sides and then use the rule $\ln(x^n) = n\ln(x)$ to bring the exponent down.

$$7 = 4^x \implies \ln(7) = \ln(4^x)$$
$$\implies \ln(7) = x\ln(4)$$
$$\implies x = \frac{\ln(7)}{\ln(4)} \approx 1.403677$$

Of course, one could use logarithms of any base, but base e is probably the most obvious choice, with the second most obvious choice being base 10. The base does not matter, you will always end up with the same numerical answer, as you can easily see.

$$7 = 4^x \implies \log(7) = \log(4^x)$$
$$\implies \log(7) = x\log(4)$$
$$\implies x = \frac{\log(7)}{\log(4)} \approx 1.403677$$

Now we continue by considering the graphs of logarithmic functions.

Example 3.8

Graph the logarithmic functions $f(x) = \log_2(x)$, $f(x) = \log_3(x)$, and $f(x) = \log_4(x)$.

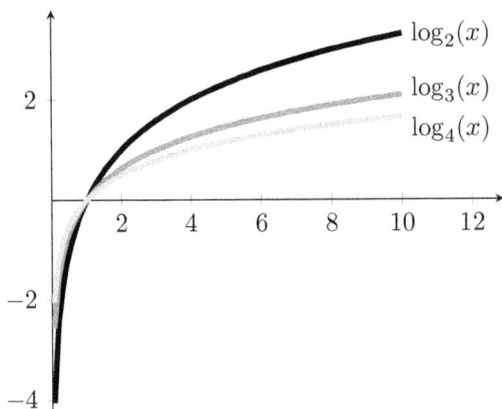

Look carefully at the graphs of the logarithmic functions. The graphs of the logarithmic functions $f(x) = \log_b(x)$, where $b > 1$, all look very similar. They all have the following properties:

1. The function does not exist if $x < 0$.
2. The function is always increasing.
3. When $x = 1$ then $f(x) = 0$. (Since any number raised to the zeroth power is one.)
4. The derivative at any point is positive.
5. The slope of the tangent line to the graph of the function at any given point is positive.

Notice, having a base of 1 for a logarithmic function does not make much sense.

Example 3.9

Find $f(x) = \log_1(5)$.

Solution: What we are looking for is the y value such that $1^y = 5$. But we already know that 1 raised to any power is equal to 1, so there is no number y that could possibly make the equation $1^y = 5$ true. Thus the question does not make sense.

But it is possible to have a base that is between zero and one.

Example 3.10

Graph the logarithmic functions $f(x) = \log_{\frac{1}{2}}(x)$, $f(x) = \log_{\frac{1}{3}}(x)$, and $f(x) = \log_{\frac{1}{4}}(x)$.

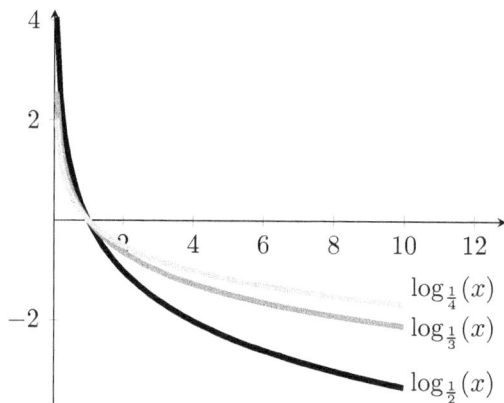

$\log_{\frac{1}{4}}(x)$

$\log_{\frac{1}{3}}(x)$

$\log_{\frac{1}{2}}(x)$

The graphs of the logarithmic functions $f(x) = \log_b(x)$, where $0 < b < 1$, all look very similar. They all have the following properties:

1. The function does not exist if $x < 0$.
2. The function is always decreasing.
3. When $x = 1$ then $f(x) = 0$. (Since any number raised to the zeroth power is one.)
4. The derivative at any point is negative.
5. The slope of the tangent line to the graph of the function at any given point is negative.

Example 3.11

Compare the graphs of the logarithmic functions $f(x) = \log_3(x)$ and $f(x) = \log_{\frac{1}{3}}(x)$.

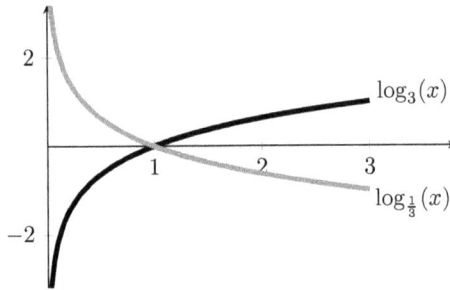

Notice these two graphs are similar. They are the reflections of each other across the x-axis.

Example 3.12

Graph the exponential function $f(x) = e^x$ and the natural logarithm function $f(x) = \ln(x)$.

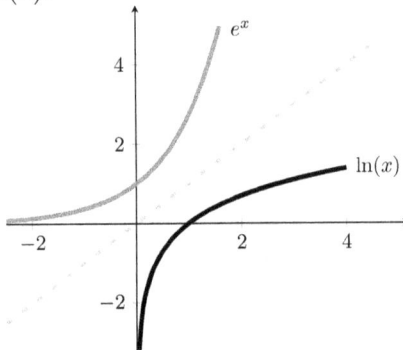

The functions are reflections of each other across the diagonal line.

The derivative of the natural logarithm function is not quite as nice as the derivative of $f(x) = e^x$, but it is still quite easy. We will not prove this fact.

The derivative of the natural logarithm function $f(x) = \ln(x)$ is given by

$$f'(x) = \frac{1}{x}.$$

3.3 DERIVATIVES OF EXPONENTIAL AND LOGARITHMIC FUNCTIONS

We are now ready to state what the derivatives of general exponential and logarithmic functions are. We will not attempt to prove those rules here.

The derivative of the general exponential functions $f(x) = b^x$, where $b > 0$, is given by

$$f'(x) = \ln(b) \cdot b^x.$$

Example 3.13

Find the derivative of $f(x) = 5^x$.

Solution: The derivative is given by

$$f'(x) = \ln(5) \cdot 5^x \approx 1.60943 \cdot 5^x.$$

We should note that most of the time people simply leave the coefficient in terms of the natural logarithm and do not give a numerical approximation for it.

Example 3.14

Find the derivative of $f(x) = \left(\frac{1}{4}\right)^x$.

Solution:

$$f'(x) = \ln\left(\frac{1}{4}\right) \cdot \left(\frac{1}{4}\right)^x$$

Example 3.15

Find the derivative of the function $f(x) = 16^x + 20$.

Solution: $$f'(x) = \ln(16)\, 16^x$$

Example 3.16

Find the derivative of the function $f(x) = 18^x - 2e^x$.

Solution: $$f'(x) = \ln(18)\, 18^x - 2e^x$$

Example 3.17

Find the derivative of the function $f(x) = \ln(x) - 5$.

Solution: $$f'(x) = \frac{1}{x}$$

Example 3.18

Find the derivative of the function $f(x) = 7\ln(x) + 10x$.

Solution: $$f'(x) = \frac{7}{x} + 10$$

Example 3.19

Find the derivative of the function $f(x) = -6e^x + x^5 - 5^x$.

Solution: $$f'(x) = -6e^x + 5x^4 - \ln(5)\, 5^x$$

Now we give the rule for the derivative of logarithmic functions.

The derivative of the general logarithmic functions $f(x) = \log_b(x)$, **where $b > 0$ and $b \neq 1$, is given by**

$$f'(x) = \frac{1}{\ln(b)} \cdot \frac{1}{x} = \frac{1}{\ln(b) \cdot x}.$$

Example 3.20

Find the derivative of $f(x) = \log_4(x)$.

Solution: The derivative is given by

$$f'(x) = \frac{1}{\ln(4)x} \approx \frac{0.72135}{x}.$$

Again, note that usually people leave the answer in terms of the natural logarithm and do not give a numerical approximation.

Example 3.21

Find the derivative of $f(x) = \log_{\frac{1}{4}}(x)$.

Solution: The derivative is given by

$$f'(x) = \frac{1}{\ln\left(\frac{1}{4}\right)x} = \frac{-1}{\ln(4)x}$$

Here we use that

$$\ln\left(\frac{1}{4}\right) = \ln\left(4^{-1}\right) = -\ln(4).$$

Example 3.22

Find the derivative of the function $f(x) = 5(7^x) - 9\log_3(x)$.

Solution: The derivative is given by

$$f'(x) = 5\ln(7)\,7^x - \frac{9}{\ln(3)x}$$

3.4 EXPONENTIAL AND LOGARITHMIC FUNCTIONS IN BUSINESS APPLICATIONS

Exponential functions are very important for a variety of real-world problems, especially problems that involve growth or decay. You can think of decay as negative growth. The function below is usually called a **growth function**.

Growth Function: Suppose some initial quantity Q_0 grows at a rate of $r\%$ per time period. After t time periods the quantity is given by

$$Q(t) = Q_0 \left(1 + \frac{r}{100}\right)^t.$$

This formula is usually given as

$$Q(t) = Q_0 \left(1 + r\right)^t$$

where it is simply understood the r in the formula is the decimal equivalent of $r\%$. If the initial quantity grows continuously at a rate of $r\%$ per time period, then after t time periods the quantity is given by

$$Q(t) = Q_0 \, e^{\frac{r}{100} \cdot t}.$$

Again, this formula is usually given as

$$Q(t) = Q_0 \, e^{r \cdot t}$$

where it is understood that the rate r in the formula is the decimal equivalent of $r\%$. In most applications, the independent variable for growth is time, hence t is generally used instead of x.

The case of continuous growth will be discussed more below, in the context of continuous interest.

Example 3.23

The population of a given city at the start of 2010 was found to be 0.75 million people. The population was estimated to grow at 0.5% per year. Estimate the population at the start of 2020. What is the instantaneous rate of change in the city's population at the start of 2020?

Solution: Since the population grows at 0.5% a year the population (in millions) after t years is given by

$$P(t) = 0.75 \left(1 + \frac{0.5}{100}\right)^t = 0.75 \, (1.005)^t.$$

From the start of 2010 to the start of 2020 is $2020 - 2010 = 10$ years, so the population at the start of 2020 is given by

$$P(10) = 0.75 \, (1.005)^{10} \approx 0.788355,$$

so the population is approximately 0.788355 million, or 788,355. To find the instantaneous rate of change in the population at the start of 2020 we need to first find the derivative of $P(t)$,

$$P'(t) = 0.75 \ \ln(1.005) \ 1.005^t.$$

Next, we evaluate the derivative at $t = 10$,

$$P'(10) = 0.75 \ \ln(1.005) \ 1.005^{10} \approx 0.003932.$$

Thus, the instantaneous rate of change of the city's population at the start of 2020 is 0.003932 million people per year or 3932 people per year.

Example 3.24

The growth of the smartphone market can be modeled by an exponential function. Assume the market is growing at a rate of 7% per year. If the current market size is 1.2 billion units, the size after t years can be modeled by $M(t) = 1.2e^{0.07t}$. Determine the market size after 5 years.

Solution:

$$M(5) = 1.2e^{0.07(5)}$$
$$= 1.2e^{0.35}$$
$$= 1.2(1.4190675)$$
$$= 1.702881 \text{ billion units.}$$

The smartphone market is expected to grow to approximately 1.7 billion units in 5 years.

Logarithmic Growth Functions

There is a certain kind of function, called a **logarithmic growth function**, that appears in business and economics applications quite often.

Logarithmic growth functions: Given constants L, C, and k, the logarithmic growth function is given by

$$N(t) = \frac{L}{1 + Ce^{-kt}}.$$

The graph of

$$N(t) = \frac{1}{1 + 100e^{-1t}}$$

is shown below.

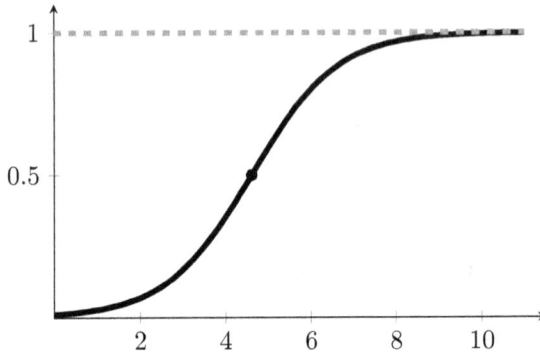

The value L determines the horizontal asymptote or the overall "height" of the function. The point of **diminishing returns** occurs at

$$t = \frac{\ln(C)}{k}.$$

The concept of diminishing returns is discussed in greater detail in Chapter 5, but in essence the point of diminishing returns occurs when the instantaneous rate of change of the function stops increasing and starts decreasing. The height of the function at this point is $\frac{L}{2}$. After Chapter 5, we will have the tools to find the point of diminishing returns for a function directly. The point of diminishing returns on the graph below is at $\left(\frac{\ln(C)}{k}, \frac{L}{2}\right)$. Also, it is important to realize that the point of diminishing returns also occurs when the growth is fastest.

Example 3.25

Suppose the spread of a computer virus is modeled by the logarithmic growth function

$$N(t) = \frac{7000}{1 + 1500e^{-1.1t}}.$$

where t is time in days and $N(t)$ is the number of computers infected. What is the maximum number of computers infected by this virus? When is the growth rate of infected computers fastest? How many computers are infected after four days?

Solution: The function is graphed below, with a point indicating the point of diminishing returns, which occurs at

$$\left(\frac{\ln(C)}{k}, \frac{L}{2}\right) \approx (6.65, 3500)$$

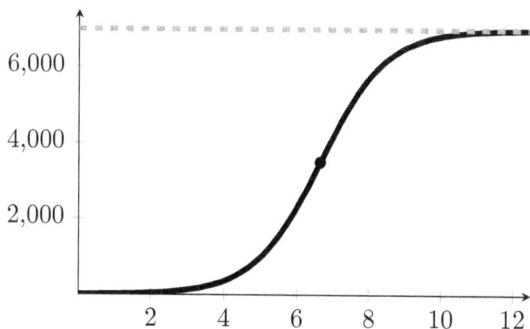

The maximum number of computers infected is simply given by $L = 7000$. The point of diminishing returns can be used to determine when the growth rate of infections is fastest, at time $t = \frac{\ln(1500)}{1.1} \approx 6.65$ days. At this point there are 3500 computers infected. To know the number of computers infected after 4 days simply evaluate the function at 4,

$$N(4) = \frac{7000}{1 + 1500e^{-1.1(4)}} = 360.52 \approx 361.$$

Example 3.26

A logistics growth function $N(t)$ is used to model the number of computers infected by a computer virus after t hours is given by

$$N(t) = \frac{83,000}{1 + 1000e^{-0.47t}}.$$

Find how many computers were affected initially (at $t = 0$). How many computers were infected overall? How many hours after the initial release of the virus was the rate of infection the greatest? And how many computers were infected at this time?

Solution: In order to tell how many computers were initially infected we evaluate the function at $t = 0$,

$$N(0) = \frac{83,000}{1 + 1000e^{-0.47(0)}} = 82.917 \approx 83.$$

Overall there was a total of $L = 83,000$ computers infected. The time at which the rate of infection was greatest is at the point of inflection,

$$t = \frac{\ln(1000)}{0.47} \approx 14.7.$$

The number of computers infected at this time was

$$\frac{83,000}{2} = 41,500.$$

Example 3.27

Consider a scenario where a company introduces a new technology product. The market penetration, or the proportion of the potential market that adopts the product over time, can often be modeled by a logarithmic growth function. Suppose the potential market size is 1,000,000 users, and the market penetration over time t (in months) follows the function:

$$N(t) = \frac{1,000,000}{1 + 500e^{-0.2t}}.$$

Analyze the market penetration after 6, 12, and 24 months. Determine the time it takes to reach 50% market penetration.

Solution: Evaluating the function at different time points gives:

- At $t = 6$ months: $N(6) = \dfrac{1,000,000}{1 + 500e^{-0.2(6)}} \approx 6596$ users.

- At $t = 12$ months: $N(12) = \dfrac{1,000,000}{1 + 500e^{-0.2(12)}} \approx 21571$ users.

- At $t = 24$ months: $N(24) = \dfrac{1,000,000}{1 + 500e^{-0.2(24)}} \approx 195508$ users.

For 50% market penetration, we solve

$$500,000 = \frac{1,000,000}{1 + 500e^{-0.2t}}$$

$$\implies \frac{1}{2} = \frac{1}{1 + 500e^{-0.2t}}$$

$$\implies 1 + 500e^{-0.2t} = 2$$

$$\implies e^{-0.2t} = \frac{1}{500}$$

$$\implies -0.2t = \ln\left(\frac{1}{500}\right)$$

$$\implies t = \frac{-\ln\left(\frac{1}{500}\right)}{0.2} = \frac{\ln(500)}{0.2} \approx 31.07.$$

Therefore, it will take 31 months to reach 50% market penetration. Notice that this is exactly the t value of the point of diminishing returns.

Interest Rate Applications

In business, the exponential function is used in solving a wide range of compound interest problems. Suppose some initial amount of P_0 was invested at an annual interest rate of $r\%$ per year, which is compounded m times a year. A logarithmic growth formula gives us the amount after t years.

Compound Interest Rate Formula: The amount of money accrued after t years from investing a principal deposit of P_0 at an annual interest rate of $r\%$ compound m times per year is giving by

$$A(t) = P_0 \left(1 + \frac{r}{m}\right)^{mt}.$$

Note, when inputting the interest rate into the formula, one must use the decimal equivalent.

The number of compoundings per year one could encounter is listed in the below table, though in practice one rarely encounters anything other than annually, quarterly, or monthly.

Compounded	m
annually	1
semi-annually	2
quarterly	4
monthly	12
weekly	52
daily	365 or 365.25

Example 3.28

You deposit $2000 in an account earning 2% compound quarterly. What is the amount in your account after 5 years?

Solution: Letting $P_0 = \$2000$, $r = 2\%$, $m = 4$, and $t = 5$, we have

$$A(t) = P_0 \left(1 + \frac{r}{m}\right)^{mt}$$

$$\implies A(t) = 2000 \left(1 + \frac{0.02}{4}\right)^{4 \cdot 5}$$

$$\implies A(t) = 2000(1+0.005)^{20}$$
$$\implies A(t) = 2000(1.005)^{20}$$
$$\implies A(t) = 2209.79.$$

Thus, a deposit of $2000 in an account earning 2% compound quarterly will result in $2209.79 after 5 years. Notice, when we substituted the 2% into the equation we used the decimal equivalent form, or 0.02.

There is one other case that is of significant theoretical interest, **continuous compounding**. Continuous compounding happens when the time interval between each compounding goes to zero. In other words, as $m \to \infty$. Recall our definition for the number e,

$$e = \lim_{n \to \infty} \left(1 + \frac{1}{n}\right)^n.$$

and notice how similar it is to our compound interest rate formula,

$$A(t) = P_0 \left(1 + \frac{r}{m}\right)^{mt}.$$

We are interested in what happens to this as $m \to \infty$. We can do some algebraic manipulations to get the following:

$$A(t) = P_0 \left(1 + \frac{r}{m}\right)^{mt}$$
$$= P_0 \left(1 + \frac{1}{m/r}\right)^{\frac{m}{r} \cdot rt}$$
$$= P_0 \left[\left(1 + \frac{1}{m/r}\right)^{m/r}\right]^{rt}$$

When $m \to \infty$ then clearly we also have $m/r \to \infty$ as well. Thus, as our number of compounding periods goes to infinity the portion of the equation in the square brackets becomes none other than the number e, and we obtain the following formula:

$$A(t) = P_0 e^{rt}.$$

This is the theoretical maximum amount one could obtain given a particular interest rate of $r\%$.

Continuously Compounded Interest Rate Formula: The amount of money accrued after t years from investing a principal deposit of P_0 at an annual interest rate of $r\%$ compounded continuously is giving by:

$$A(t) = P_0 e^{rt}.$$

Note, when inputting the interest rate into the formula, one must use the decimal equivalent.

Example 3.29

You deposit \$1500 in an account earning 1.5% compound continuously. What is the amount in your account after 10 years?

Solution: Letting $P_0 = \$1500$, $r = 1.5\%$, we have

$$A(t) = P_0 e^{rt} = 1500 e^{0.015t}$$

Then, substituting $t = 10$ in this last equation, we will find:

$$A(10) = 1500 e^{0.015(10)} = 1742.75$$

Thus, a deposit of \$1500 in an account earning 1.5% compounded continuously will result in \$1742.75 after 10 years.

Before delving into interest rate applications of exponential functions, we will provide the formula for simple interest. The simple interest formula is not actually an exponential function, but it does get used in certain real-world situations so it is important to know.

Simple Interest Rate Formula: The amount of money accrued after t years from investing a principal deposit of P_0 at an annual simple interest rate of $r\%$ is given by

$$A(t) = P_0(1 + rt).$$

Note, when inputting the interest rate into the formula, one must use the decimal equivalent. This means that the actual interest earned after t years is given by

$$I(t) = P_0 rt.$$

Example 3.30

You deposit \$1000 for 5 years into an account that pays 1.75% interest. How much have you accrued after 5 years assuming the interest rate is:
- simple,
- compounded annually,
- compounded semi-annually,
- compounded monthly,
- compounded weekly,
- compounded continuously.

Solution:

- At a simple interest rate, we have

$$A(t) = 1000\big(1 + (0.0175)5\big) = \$1087.50.$$

- When compounded annually, we have

$$A(t) = 1000 \left(1 + \frac{0.0175}{1}\right)^{1 \cdot 5} = \$1090.62.$$

- When compounded semi-annually, we have

$$A(t) = 1000 \left(1 + \frac{0.0175}{2}\right)^{2 \cdot 5} = \$1091.03.$$

- When compounded monthly, we have

$$A(t) = 1000 \left(1 + \frac{0.0175}{12}\right)^{12 \cdot 5} = \$1091.37.$$

- When compounded weekly, we have

$$A(t) = 1000 \left(1 + \frac{0.0175}{52}\right)^{52 \cdot 5} = \$1091.43.$$

- When compounded continuously, we have

$$A(t) = 1000e^{0.0175 \cdot 5} = \$1091.44.$$

Since we are talking about money we rounded to the nearest cent. Notice that as the number of compounding increased the amount of money earned in interest also increased. Since continuous compounding essentially represents an infinite number of compounding periods, this is the theoretical maximum amount one could earn at a given interest rate. Thus, continuous compounding is of theoretical interest. (No pun intended!)

It is of course possible to ask a wide variety of questions in interest rate problems. For example, the unknown quantity could be the initial amount, the interest rate, or the time. Generally, we do not assume the number of compoundings is unknown.

Example 3.31

A child's parents want to save for college. They can put their money in a savings account that pays 3.5% compounded quarterly. How much money would they need to put into the account to have $100,000 in 15 years?

Solution: Here the unknown is the initial amount P_0.

$$100,000 = P_0 \left(1 + \frac{0.035}{4}\right)^{4 \cdot 15}$$

$$= P_0 \, (1.00875)^{60}$$

$$\implies P_0 = \frac{100,000}{(1.00875)^{60}} = \$59,290.78.$$

An amount of $\$59,290.78$ needs to be invested now at an interest rate of 3.5% compounded quarterly in order to have $\$100,000$ in 15 years.

Example 3.32

An individual wants to have $\$25,000$ in four years. She currently has $\$19,500$ in a saving account. Assuming the interest is compounded quarterly, what interest rate is needed?

Solution: Here the unknown is the interest rate r.

$$25000 = 19500 \left(1 + \frac{r}{4}\right)^{4 \cdot 4}$$

$$\implies \frac{25000}{19500} = \left(1 + \frac{r}{4}\right)^{16}$$

$$\implies \left(\frac{25000}{19500}\right)^{\frac{1}{16}} = 1 + \frac{r}{4}$$

$$\implies \sqrt[16]{\frac{25000}{19500}} - 1 = \frac{r}{4}$$

$$\implies r = 4 \cdot \left[\sqrt[16]{\frac{25000}{19500}} - 1\right] = 0.0626.$$

An interest rate of 6.26% compounded quarterly is needed to have $\$25,000$ in four years.

Example 3.33

Suppose someone has $\$5,000$ in an account that pays 3% compounded semi-annually. How long does it take to double their money?

Solution: Here the unknown is the number of years t and double $\$5,000$ is $\$10,000$, so

$$10000 = 5000 \left(1 + \frac{0.03}{2}\right)^{2 \cdot t}$$

$$\implies \frac{10000}{5000} = \left((1.015)^2\right)^t$$

$$\implies \ln(2) = \ln\left(1.015^2\right)^t$$

$$\implies \ln(2) = t \cdot \ln\left(1.015^2\right)$$

$$\implies t = \frac{\ln(2)}{\ln(1.015^2)} = 23.28.$$

In the third line, we took the natural log of both sides of the equation and in the fourth line, we used a logarithm property to bring the exponent t down. The money doubles in 23.28 years.

Example 3.34

Suppose \$2000 is invested in an account that compounds interest continuously at a rate of 5.5%. How many years will it take for the amount in the account to be \$3000?

Solution: Here the unknown is t,

$$3000 = 2000e^{0.055t}$$

$$\implies \frac{3000}{2000} = e^{0.055t}$$

$$\implies \ln(1.5) = \ln\left(e^{0.055t}\right)$$

$$\implies \ln(1.5) = t \cdot \ln\left(e^{0.055}\right)$$

$$\implies \ln(1.5) = t \cdot 0.055$$

$$\implies t = \frac{\ln(1.5)}{0.055} = 7.37.$$

In the third line, we took the natural log of both sides, in the fourth line, we used a property of logarithms to bring the t down, and in the fifth line, we used the fact that ln and e are inverse functions of each other. It will take 7.37, or approximately 7 years and four months, in order for the amount to reach \$3000.

One of the main types of problems encountered in practice is choosing between two or more different investment opportunities. One criteria that is often used in deciding which investment to choose is the amount of return. We will use this criteria here. Other criteria may take into account the riskiness of the investment; we will not consider that.

Example 3.35

Suppose you have \$10,000 to invest and can choose between the following two investment opportunities. Which opportunity will provide the greatest return?

- An account that pays 1.9% compounded annually.
- An account that pays 1.85% compounded weekly.

Solution: In general, the higher the interest rate, the greater the return. But also, the higher the number of compoundings, the greater the return. Here we are offered a choice between one investment that has a higher rate but fewer compoundings, and another investment that has a lower rate but more compoundings. How do we choose between these two options? One possibility is to simply assume a one-year investment period and calculate the expected accrued amounts.

- When an account pays 1.9% compounded annually after one year, we have
$$A(1) = 10,000 \left(1 + \frac{0.019}{1}\right)^{1 \cdot 1} = \$10,190.00.$$

- An account pays 1.85% compounded weekly; after one year, we have
$$A(1) = 10,000 \left(1 + \frac{0.0185}{52}\right)^{52 \cdot 1} = \$10,186.69.$$

After one year, the accrued amount from the first option is slightly higher than the second option. Thus, we would choose the first option.

Consider the second case, where we earned 1.85% compounded weekly. Let us redo the computation, but this time with $P_0 = 1$,

$$A(1) = 1 \left(1 + \frac{0.0185}{52}\right)^{52 \cdot 1} = \$1.018669.$$

It is easy to see that this is exactly the same amount that we would have obtained if we had an interest rate of 1.8669% compounded annually;

$$A(1) = 1 \left(1 + \frac{0.018669}{1}\right)^{1 \cdot 1} = \$1.08669.$$

This gives us the concept of an **effective interest rate**. The *effective interest rate* on an investment with an interest rate of 1.85% compounded weekly is 1.8669% (compounded annually). In other words, the effective interest rate is the interest rate we would need to have to obtain the same amount of money if there is only one compounding per year. Interest rates that are not effective interest rates are sometimes called **nominal interest rates**. We can use the concept of effective interest rates to compare different investment opportunities. We would calculate the effective interest rate of each opportunity and choose the option with the highest effective rate.

Effective Interest Rate Formulas: Given a (nominal) interest rate of $r\%$ compounded m times per year, the effective interest rate is given by

$$r_{\text{eff}} = \left(1 + \frac{r}{m}\right)^m - 1.$$

Given a (nominal) interest rate of $r\%$ compounded continuously, the effective interest rate is given by

$$r_{\text{eff}} = e^r - 1.$$

Example 3.36

Suppose we have one investment opportunity that pays 8% compounded quarterly and another that pays 7.75% compounded continuously. Use the effective interest rate to determine which opportunity pays more.

Solution: For the first investment, we have

$$r_{\text{eff}} = \left(1 + \frac{0.08}{4}\right)^4 - 1 = 0.08243,$$

or 8.243%. For the second investment, we have

$$r_{\text{eff}} = e^{0.0775} - 1 = 0.08058,$$

or 8.058%. Based on the effective interest rates, the first investment will pay more.

Interest rates play a crucial role in determining the cost of borrowing. They significantly affect the monthly repayment amounts for loans, as higher rates increase the repayment burden.

Loan Repayment Formula: The monthly repayment, denoted by PMT, over t years can be calculated using the formula

$$PMT = P \cdot \frac{r(1+r)^n}{(1+r)^n - 1},$$

where P is the principal, r is the monthly interest rate, and n is the total number of monthly payments, given by $n = 12t$.

Example 3.37

For a loan of $10,000 at an annual interest rate of 5%, compounded monthly, calculate the monthly repayment over 5 years.

Solution: First we start by converting the annual interest rate to a monthly rate by dividing it by 12 (months in a year): $r = \frac{0.05}{12} \approx 0.004167$. Then, the total number of payments n over 5 years, with monthly payments, is $5 \times 12 = 60$.

$$PMT = 10{,}000 \cdot \frac{0.004167\left(1 + 0.004167\right)^{60}}{\left(1 + 0.004167\right)^{60} - 1}$$

$$= 10{,}000 \cdot \frac{0.004167(1.28304)}{1.28304 - 1}$$

$$= \$188.71.$$

The monthly repayment amount for the loan is approximately $188.71.

Logarithmic functions are very important in econometrics, especially in log-linear models and stock market analysis. In log-linear models variables are transformed using the natural logarithm, this is particularly useful in regression analysis where relationships between economic variables are not strictly linear. Furthermore, in stock market analysis, logarithmic functions are essential for calculating **continuously compounded returns**, which are also called **logarithmic returns**. Continuously Compounded Returns, often denoted by r_{cc}, is a financial concept used to calculate the return on an investment (or loan) where we assume the interest is compounded continuously over time.

The formula itself is a direct consequence of the continuously compounded interest rate formula, $A(t) = P_0 e^{rt}$. Suppose, for example, the value of a stock was P_t at the start of a time period, and P_{t+1} at the end of the time period. Letting $P_0 = P_t$ and $A(1) = P_{t+1}$, we have

$$P_{t+1} = P_t e^{r \cdot 1}$$

$$\implies \frac{P_{t+1}}{P_t} = e^r$$

$$\implies r = \ln\left(\frac{P_{t+1}}{P_t}\right).$$

The continuously compounded interest rate r, which is often denoted by r_{cc}, is the continuously compunded interest rate earned by that stock over the time period.

Continuous Compound Return: The formula often used in finance and economics to calculate the logarithmic return (or log return) of an investment is given as

$$r_{cc} = \ln\left(\frac{P_{t+1}}{P_t}\right)$$

where P_t is the investment value at time t, and P_{t+1} is the investment value at time $t + 1$.

Example 3.38

Suppose you have a stock market that was priced at 200 at time t and then rose to 205 at time $t + 1$. Find the continuously compound return over this period.

Solution: To calculate the continuous compound return over this period, we use the formula given above as follows:

$$r_{cc} = \ln\left(\frac{P_{t+1}}{P_t}\right) = \ln\left(\frac{205}{200}\right) = \ln(1.025) = 0.0247.$$

Therefore, the continuous compound return is approximately 2.47%. In other words, if you invest $200, compounded continuously at 2.47% a year, you will have $205 in one year.

This logarithmic calculation of continuous compound return is critical in finance because it allows for the adjustment of returns for different investments to a standard rate, facilitating more accurate comparisons.

More Applications

Example 3.39

If the cost of producing x thousand units is given by the logarithmic function $C(x) = 500 + 100\ln(x)$ and the revenue is $R(x) = 150x$, find the break-even point.

Solution: the break-even point happened when $C(x) = R(x)$, that is, when

$$500 + 100\ln(x) = 150x.$$

Solving this equation algebraically can be challenging due to the logarithmic term. Such equations often don't have simple algebraic solutions. In cases where finding an exact algebraic solution is difficult, graphing can provide a good approximation.

As shown in the graph, the break-even point occurs approximately at $x = 4.26$ thousand units. This means that producing and selling about 4260 units will result in neither profit nor loss.

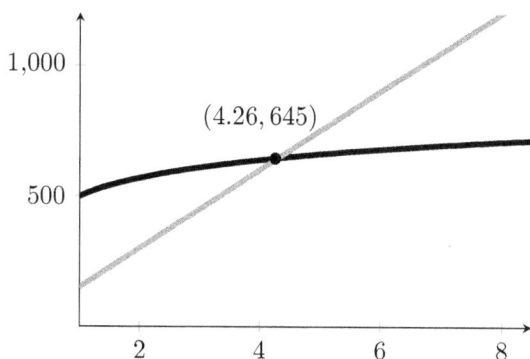

Example 3.40

Suppose the cost $C(x)$ of producing x units of a product is given by a logarithmic function, $C(x) = 500 + 150 \ln(x)$. Determine the marginal cost when $x = 100$ units.

Solution: The marginal cost, representing the cost of producing one additional unit, is the derivative $MC(x) = C'(x)$,

$$MC(x) = \left(500 + 150 \ln(x)\right)' = \frac{150}{x}.$$

Then we substitute 100 in the equations to find $MC(100) = \frac{150}{100} = 1.5$. Thus, the approximate cost of producing the 101st unit is \$1.50.

Example 3.41

A librarian used data from the university library to build the following model:

$$A(t) = 120 \ln(t) + 32,$$

where $A(t)$ is the number of library borrowers t years after 2020.

1. Estimate the number of borrowers in 2024 and interpret the result.
2. How rapidly is the number of borrowers increasing per year?

Solution:

1. To estimate the number of borrowers in 2024, we substitute $t = 4$ into the model:

$$A(4) = 120 \ln(4) + 32 \approx 198.36.$$

Therefore, the estimated number of borrowers in 2024 is approximately 198.

2. To find the rate at which the number of borrowers is increasing, we take the derivative of $A(t)$ with respect to t:

$$A'(t) = \left(120 \ln(t) + 32\right)' = \frac{120}{t}.$$

Evaluating this at $t = 4$, we find the rate of change of borrowers:

$$A'(4) = \frac{120}{4} = 30.$$

Thus, in 2024, the number of borrowers is increasing at a rate of 30 borrowers per year.

Example 3.42

A store sells used books. If the store sells x books at a price of p dollars per book, then the price demand equation is $p = 400(0.899)^x$. Find the rate of change of price with respect to demand when the demand is 50 books and interpret the answer.

Solution: To find the rate of change of price with respect to demand we calculate the derivative of p with respect to x,

$$p' = 400 \cdot \ln(0.899) \cdot (0.899)^x.$$

We then evaluate this derivative at $x = 50$,

$$p' = 400 \cdot \ln(0.899) \cdot (0.899)^{50} \approx -0.21.$$

The rate of change of price with respect to demand at 50 books is approximately -0.21 dollars per book. In other words, if we decrease the price of the books by 21 cents, we would expect the demand to increase from 50 books to 51 books.

Example 3.43

Assume the demand $D(p)$ for a product in a competitive market is given by the exponential function $D(p) = 100e^{-0.05p}$, where p is the price per unit. What is the rate of change of demand with respect to price? Interpret the meaning of the derivative in terms of demand and price. Consider what a negative value for the rate of change implies about the relationship between demand and price.

Solution: To find the rate of change of demand with respect to price, we calculate the derivative of the demand function with respect to p:

$$D'(p) = (100e^{-0.05p})'$$

$$= 100 \cdot \left(e^{-0.05p}\right)'$$

$$= 100 \cdot \left((e^{-0.05})^p\right)'$$

$$= 100 \cdot \ln(e^{-0.05}) \cdot e^{-0.05p}$$

$$= 100 \cdot (-0.05) \cdot e^{-0.05p}$$

$$= -5e^{-0.05p}.$$

Note, in line 3, we used the exponent property to write $e^{-0.05p} = (e^{-0.05})^p$, in line 4, we used the derivative of b^p with $b = e^{-0.05}$, and in line 5 we used that $\ln$ and e are inverse functions to write $\ln(e^{-0.05}) = -0.05$.

The derivative of the demand function with respect to the price p, given by $-5e^{-0.05p}$, represents the rate of change of demand in response to changes in price. A negative value for this rate of change implies that there is an inverse relationship between price and demand, which is a fundamental concept in economics known as the law of demand. Specifically, the rate of change $D'(p)$ tells us that for a small increase in the price, the quantity of the product demanded decreases. The exponential decay factor $e^{-0.05p}$ indicates that this rate of change will decrease as the price increases.

3.5 PROBLEMS

Question 3.1 *Find the derivatives of the following functions:*

(a) $f(x) = \ln(x) + 5$

(b) $f(x) = 15\ln(x) - 7$

(c) $f(x) = 8\log_2(x)$

(d) $f(x) = 10(\log_7(x)) + x^2$

(e) $f(x) = 30^x$

(f) $f(x) = e^x + 10x - 20$

(g) $f(x) = 3(2^x) - 4e^x$

(h) $f(x) = \log_{\frac{1}{7}}(x)$

(i) $f(x) = \left(\frac{3}{5}\right)^x$

Question 3.2 *After how many years an initial deposit of $6,000 invested at 3% interest rate compound annually, will double?*

Question 3.3 *A firm could like to invest $100,000 for a period of 20 years. The firm has two options: 3% interest rate compound every 6 months or 2.5% interest rate compound every 2 months. Which option should the firm opt for?*

Question 3.4 *What initial deposit should you consider for an investment at an interest rate of 4% compound annually, to have an amount if $9,000 after 6 years?*

Question 3.5 *In order the find a relationship between weight and diastolic blood pressure in youth aged 5 to 19 years, a researcher used data from the city hospital. The relationship based on 2089 records was approximated by the following function:*

$$f(x) = 74.5(2 + \log(x))$$

where $f(x)$ is the diastolic blood pressure measured in millimeters of mercury (mmHg) and x is the weight in kilograms (kg).
   1. *What is the rate of change of the diastolic blood pressure with respect to weight at the 63kg weight?*

   2. *What is the weight at which the rate of change of the diastolic blood pressure with respect to weight is 0.5 mmHg per kg?*

Question 3.6 *Determine if each of the following functions is an example of exponential growth or decay.*

   1. $f(x) = \frac{1}{3}4^x$.
   2. $f(x) = 3(0.2)^x$.
   3. $f(x) = 7(2.3)^x$.

Question 3.7 *A specific bacteria population is known to double every 200 minutes. If there are initially 60 bacteria. Write a function that models the size of this bacteria after t hours.*

Question 3.8 *Let $C(t) = 200e^{(-0.8t)}$ be the concentration of a certain drug in the bloodstream, measured in micrograms per milliliter, t hours after the drug is administered. What is the rate of change of concentration after 2 hours? after 5 hours?*

Question 3.9 *The population of Miracle City at the start of 2020 was found to be 1.3 million people. The population was estimated to grow at 0.75% per year. Estimate the population at the start of 2030. What is the instantaneous rate of change in the city's population at the start of 2030?*

Question 3.10 *An epidemic spreads among the population following the following logistic growth model:*

$$N(t) = \frac{10}{1 + 9e^{-0.83t}}$$

where t is the number of days. Estimate the number of cases of this pandemic after 35 days.

Question 3.11 *The logistic growth model of a certain bacteria is given by the following model:*

$$N(t) = \frac{12}{1 + 11e^{-t}}$$

where t is the number of years. What was the bacteria population at the beginning and what is the rate of change of the bacteria population after 4 years?

Question 3.12 *For a loan of $8,000 at an annual interest rate of 6%, compounded monthly, calculate the monthly repayment over a period of 3 years.*

Question 3.13 *Suppose you have invested $10,000 in a savings account that offers a continuously compounded interest rate. If the value of the investment rises to $10,500 after one year, find the continuously compounded annual interest rate for this period.*

Question 3.14 *Given the market growth model $M(t) = 1.2 \times e^{0.07t}$ for the smartphone market, where t is in years, calculate the market size after 3 years.*

Question 3.15 *Given the revenue function $R(x) = 5x + e^{0.01x}$ for a product, determine the revenue when 150 units are sold.*

Question 3.16 *For the cost function $C(x) = 1000 + 200\ln(x)$ and the revenue function $R(x) = 500x$, where x is in thousands of units, determine the number of units that need to be sold to break even.*

Question 3.17 *A local coffee shop finds that it can sell x cups of coffee at a price of \$p dollars per cup, where the price-demand equation is given by* $p = 200 \times (0.95)^x$.

1. *Calculate the rate of change of price with respect to the number of cups sold when the sales are 100 cups.*
2. *Interpret what this rate of change means for the coffee shop's pricing strategy.*

Question 3.18 *A small community library tracks its book checkouts over time and models them with the function* $B(t) = 150 \ln(t) + 20$, *where* $B(t)$ *is the number of books checked out in the year* t, *with* $t = 0$ *representing the year 2020.*

1. *Predict the number of books that will be checked out in the year 2025 and provide an interpretation of this prediction.*
2. *Determine the rate at which book checkouts are increasing annually.*

Question 3.19 *A technology company models the demand* $D(p)$ *for its latest smartphone in the market with the function* $D(p) = 80e^{-0.1p}$, *where* p *stands for the price of the smartphone in dollars.*

1. *Find the rate of change of demand with respect to the smartphone's price.*
2. *Discuss the implications of this rate of change for the company's pricing policy.*

CHAPTER **4**

Rules of Differentiation

In this chapter, we will present you with a number of rules for finding the derivative of many functions.

4.1 THE PRODUCT RULE

The Product Rule:

Suppose we have a function $F(x)$ which is the product of two differentiable functions $f(x)$ and $g(x)$, that is, $F(x) = f(x) \cdot g(x)$. Then

$$F'(x) = f'(x)g(x) + f(x)g'(x).$$

Sometimes the product rule is written in terms of functions $u(x)$ and $v(x)$, which are simply written as u and v. Suppose $y = uv$, then

$$y' = u'v + uv'.$$

The proof of the product rule is not hard, but it does involve some algebra. The trick is to both subtract and add the term $f(x + h)g(x)$ in the second line, which is shown in bold.

$$[f(x) \cdot g(x)]' = \lim_{h \to 0} \frac{f(x+h)g(x+h) - f(x)g(x)}{h}$$

$$= \lim_{h \to 0} \frac{f(x+h)g(x+h) - \mathbf{f(x+h)g(x)} + \mathbf{f(x+h)g(x)} - f(x)g(x)}{h}$$

$$= \lim_{h \to 0} \left(\frac{f(x+h)g(x+h) - f(x+h)g(x)}{h} + \frac{f(x+h)g(x) - f(x)g(x)}{h} \right)$$

$$= \lim_{h \to 0} \left(f(x+h)\frac{g(x+h) - g(x)}{h} + g(x)\frac{f(x+h) - f(x)}{h} \right)$$

$$= \lim_{h \to 0} \left(f(x+h)\frac{g(x+h) - g(x)}{h} \right) + \lim_{h \to 0} \left(g(x)\frac{f(x+h) - f(x)}{h} \right)$$

$$= \lim_{h \to 0} f(x+h) \cdot \lim_{h \to 0} \left(\frac{g(x+h) - g(x)}{h} \right)$$

$$+ \lim_{h \to 0} g(x) \cdot \lim_{h \to 0} \left(\frac{f(x+h) - f(x)}{h} \right)$$

$$= f(x)g'(x) + g(x)f'(x)$$

$$= f'(x)g(x) + f(x)g'(x).$$

For us, far more important than the proof, is being able to use the rule correctly. We start with a few easy examples.

Example 4.1

Suppose we have $y = x^2 e^x$. Find the derivative of y.

Solution: Letting $u = x^2$ and $v = e^x$, it is easy to find $u' = 2x$ and $v' = e^x$. Applying the formula for the product rule, we have

$$y = u'v + uv'$$

$$= 2xe^x + x^2 e^x$$

$$= (x^2 + 2x)e^x.$$

In the last step we simply factored out the e^x and wrote the polynomial terms in decreasing order. Note, we could have also factored out an x.

Example 4.2

Suppose we have $y = 3x^2 \ln(x)$. Find the derivative of y.

Solution: Letting $u = 3x^2$ and $v = \ln(x)$, it is easy to find $u' = 6x$ and $v' = \frac{1}{x}$. Applying the formula for the product rule, we have

$$y = u'v + uv'$$

$$= 6x \ln(x) + 3x^2 \frac{1}{x}$$

$$= 6x \ln(x) + 3x$$

$$= 3x \Big(2 \ln(x) + 1 \Big).$$

Again, in the last step we factored out what we could, though the second to last step is also generally an acceptable answer.

Example 4.3

Suppose $F(x) = e^x \ln(x)$. Find $F'(x)$.

Solution: Letting $f(x) = e^x$ and $g(x) = \ln(x)$, we have $f'(x) = e^x$ and $g'(x) = \frac{1}{x}$. Applying the formula for the product rule, we have

$$F'(x) = f'(x)g(x) + f(x)g'(x)$$
$$= e^x \ln(x) + e^x \frac{1}{x}$$
$$= e^x \left(\ln(x) + \frac{1}{x} \right).$$

Some examples can be a bit more complicated, especially if the final answer needs to be simplified.

Example 4.4

Suppose $F(x) = (x^2 - 7x)(3x^4 + 5x^3 - 2x^2)$. Find the derivative of $F(x)$.

Solution: Here, we have $f(x) = x^2 - 7x$ and $g(x) = 3x^4 + 5x^3 - 2x^2$. This gives us $f'(x) = 2x - 7$ and $g'(x) = 12x^3 + 15x^2 - 4x$. Applying the formula for the product rule, we have

$$F'(x) = f'(x)g(x) + f(x)g'(x)$$
$$= (2x - 7)(3x^4 + 5x^3 - 2x^2) + (x^2 - 7x)(12x^3 + 15x^2 - 4x).$$

If we need to simplify this, we have a few extra steps to do, but end up with

$$
\begin{aligned}
F'(x) &= (2x - 7)(3x^4 + 5x^3 - 2x^2) + (x^2 - 7x)(12x^3 + 15x^2 - 4x) \\
&= (2x - 7)3x^4 + (2x - 7)5x^3 - (2x - 7)2x^2 \\
&\quad + (x^2 - 7x)12x^3 + (x^2 - 7x)15x^2 - (x^2 - 7x)4x \\
&= 2x(3x^4) - 7(3x^4) + 2x(5x^3) - 7(5x^3) - 2x(2x^2) + 7(2x^2) \\
&\quad + x^2(12x^3) - 7x(12x^3) + x^2(15x^2) - 7x(15x^2) - x^2(4x) + 7x(4x) \\
&= 6x^5 - 21x^4 + 10x^4 - 35x^3 - 4x^3 + 14x^2 \\
&\quad + 12x^5 - 84x^4 + 15x^4 - 105x^3 - 4x^3 + 28x^2 \\
&= 18x^5 - 80x^4 - 148x^3 + 42x^2.
\end{aligned}
$$

Example 4.5

Suppose $F(x) = (x^2 + 3x + 2)(3x^2 - 5x + 1)$. Find $F'(x)$.

Solution: Letting $f(x) = x^2 + 3x + 2$ and $g(x) = 3x^2 - 5x + 1$, we have $f'(x) = 2x + 3$ and $g'(x) = 6x - 5$. Applying the formula for the product rule, we have

$$F'(x) = (2x + 3)(3x^2 - 5x + 1) + (x^2 + 3x + 2)(6x - 5).$$

Simplifying would give us $F'(x) = 12x^3 + 12x^2 - 16x - 7$.

Example 4.6

Suppose $F(x) = (3x^3 + 4x^2)(x^{-1})$. Find $F'(x)$.

Solution: Letting $f(x) = 3x^3 + 4x^2$ and $g(x) = x^{-1}$, we have $f'(x) = 9x^2 + 8x$ and $g'(x) = -x^{-2}$. Applying the formula for the product rule, we have

$$F'(x) = (9x^2 + 8x)(x^{-1}) + (3x^3 + 4x^2)(-x^{-2}).$$

Simplifying would give us $F'(x) = 6x + 4$.

4.2 THE QUOTIENT RULE

The Quotient Rule:

Suppose we have a function $F(x)$ which is the quotient of two differentiable functions $f(x)$ and $g(x)$; that is, $F(x) = \frac{f(x)}{g(x)}$. Then

$$F'(x) = \frac{g(x)f'(x) - f(x)g'(x)}{\left[g(x)\right]^2}.$$

Sometimes the quotient rule is written in terms of functions $u(x)$ and $v(x)$, which are simply written as u and v. Suppose $y = \frac{u}{v}$, then

$$y' = \frac{vu' - uv'}{v^2}.$$

The proof of the quotient rule is not hard, but does involve some algebra. Like for the product rule, the trick is to both subtract and add the term $f(x)g(x)$

in the third row. This is shown in bold.

$$\left(\frac{f(x)}{g(x)}\right)' = \lim_{h \to 0} \frac{\frac{f(x+h)}{g(x+h)} - \frac{f(x)}{g(x)}}{h}$$

$$= \lim_{h \to 0} \frac{g(x)f(x+h) - f(x)g(x+h)}{g(x)g(x+h)h}$$

$$= \lim_{h \to 0} \frac{g(x)f(x+h) - \boldsymbol{f(x)g(x)} + \boldsymbol{f(x)g(x)} - f(x)g(x+h)}{g(x)g(x+h)h}$$

$$= \lim_{h \to 0} \frac{g(x)\left[f(x+h) - f(x)\right] + f(x)\left[g(x) - g(x+h)\right]}{g(x)g(x+h)h}$$

$$= \lim_{h \to 0} \frac{\frac{g(x)\left[f(x+h) - f(x)\right]}{h} + \frac{f(x)\left[g(x) - g(x+h)\right]}{h}}{g(x)g(x+h)}$$

$$= \frac{g(x) \cdot \lim_{h \to 0} \frac{f(x+h) - f(x)}{h} + f(x) \cdot \lim_{h \to 0} \frac{g(x) - g(x+h)}{h}}{\lim_{h \to 0} g(x)g(x+h)}$$

$$= \frac{g(x)f'(x) - f(x)g'(x)}{g(x)g(x)}$$

$$= \frac{g(x)f'(x) - f(x)g'(x)}{\left[g(x)\right]^2}.$$

Like with the product rule, the actual proof is not so important for us, but knowing how to use the quotient rule is important. We will start with a few easy examples.

Example 4.7

Suppose we have $y = \dfrac{x}{e^x}$. Find the derivative of y.

Solution: Letting $u = x$ and $v = e^x$, it is easy to see $u' = 1$ and $v' = e^x$. Applying the formula for the quotient rule, we have

$$y' = \frac{vu' - uv'}{v^2}$$

$$= \frac{e^x(1) - x(e^x)}{(e^x)^2}$$

$$= \frac{e^x(1 - x)}{(e^x)^2}$$

$$= \frac{1 - x}{e^x}.$$

Example 4.8

Suppose we have $y = \dfrac{x^2 + 1}{x - 1}$. Find the derivative of y.

Solution: Letting $u = x^2 + 1$ and $v = x - 1$, it is easy to see $u' = 2x$ and $v' = 1$. Applying the formula for the quotient rule, we have

$$
\begin{aligned}
y' &= \frac{vu' - uv'}{v^2} \\
&= \frac{(x-1)(2x) - (x^2+1)(1)}{(x-1)^2} \\
&= \frac{2x^2 - 2x - x^2 - 1}{(x-1)^2} \\
&= \frac{x^2 - 2x - 1}{(x-1)^2}.
\end{aligned}
$$

Example 4.9

Suppose $F(x) = \dfrac{\ln(x)}{x^2}$. Find $F'(x)$.

Solution: Letting $f(x) = \ln(x)$ and $g(x) = x^2$, we have $f'(x) = \frac{1}{x}$ and $g'(x) = 2x$. Applying the formula for the quotient rule, we have

$$
\begin{aligned}
F'(x) &= \frac{g(x)f'(x) - f(x)g'(x)}{[g(x)]^2} \\
&= \frac{x^2 \frac{1}{x} - \ln(x)(2x)}{(x^2)^2} \\
&= \frac{x - 2x\ln(x)}{x^4} \\
&= \frac{x(1 - 2\ln(x))}{x^4} \\
&= \frac{1 - 2\ln(x)}{x^3}.
\end{aligned}
$$

Example 4.10

Suppose $F(x) = \dfrac{-7x^3 + 5x^2}{e^x}$. Find $F'(x)$.

Solution: Letting $f(x) = -7x^3 + 5x^2$ and $g(x) = e^x$, we have $f'(x) = -21x^2 + 10x$ and $g'(x) = e^x$. Applying the formula for the quotient rule, we have

$$F'(x) = \frac{g(x)f'(x) - f(x)g'(x)}{[g(x)]^2}$$

$$= \frac{e^x(-21x^2 + 10x) - (-7x^3 + 5x^2)e^x}{(e^x)^2}$$

$$= \frac{e^x(-21x^2 + 10x + 7x^3 - 5x^2)}{(e^x)^2}$$

$$= \frac{7x^3 - 26x^2 + 10x}{e^x}.$$

We could easily, if we wanted, also factor an x out of the numerator.

Example 4.11

Suppose we have $y = \dfrac{x^3 + 4}{x^2 - 3}$. Find the derivative of y.

Solution: Letting $u = x^3 + 4$ and $v = x^2 - 3$, it is easy to find $u' = 3x^2$ and $v' = 2x$. Applying the formula for the quotient rule, we have

$$y' = \frac{(x^2 - 3)(3x^2) - (x^3 + 4)(2x)}{(x^2 - 3)^2}$$

$$= \frac{(3x^4 - 9x^2) - (2x^4 + 8x)}{(x^2 - 3)^2}$$

$$= \frac{x^4 - 9x^2 - 8x}{(x^2 - 3)^2}.$$

4.3   THE CHAIN RULE

The chain rule for derivatives applies to functions which have been composed. Recall, given two functions $f(x)$ and $g(x)$, then f composed with g is written as $f \circ g$, which means

$$(f \circ g)(x) = f\Big(g(x)\Big).$$

In other words, the function g is the input for the function f. Sometimes we may say that f is "eating" g.

The Chain Rule:

If f is differentiable at $g(x)$ and g is differentiable at x then the function $F(x) = f(g(x))$ is differentiable at x and the derivative is

$$F'(x) = f'\Big(g(x)\Big) \cdot g'(x)$$

If $y = f(u)$ and $u = g(x)$ then using Leibniz notation this is written as

$$\frac{dy}{dx} = \frac{dy}{du}\frac{du}{dx}.$$

The proof of the chain rule requires a slightly different version of the definition of the derivative. Recall, we had defined the derivative of $f(x)$ at x to be

$$f'(x) = \lim_{h \to 0} \frac{f(x+h) - f(x)}{h}.$$

Now suppose we wanted to find the derivative of $f(x)$ at the value c. We could rewrite the definition of derivative as

$$f'(c) = \lim_{x \to c} \frac{f(x) - f(c)}{x - c}.$$

Notice, this is essentially the same thing with the denominator going to zero. This is the version of the definition for the derivative we need to use in this proof. For this proof we also have to divide and multiply by the term $g(x) - g(c)$, which was done in bold. Let $h(x) = f(g(x))$. We need to show for $x = c$ that $h'(c) = f'(g(c))g'(c)$.

$$
\begin{aligned}
h'(c) &= \Big(f(g(c))\Big)' \\
&= \lim_{x \to c} \frac{f(g(x)) - f(g(c))}{x - c} \\
&= \lim_{x \to c} \left(\frac{f(g(x)) - f(g(c))}{g(x) - g(c)} \cdot \frac{g(x) - g(c)}{x - c} \right) \\
&= \left(\lim_{x \to c} \frac{f(g(x)) - f(g(c))}{g(x) - g(c)} \right) \cdot \left(\lim_{x \to c} \frac{g(x) - g(c)}{x - c} \right) \\
&= f'(g(c))g'(c).
\end{aligned}
$$

And of course, since this holds for any value c we simply replace the c with an x. Writing down the chain rule makes it look harder than it actually is. Taking derivatives using the chain rule is not any more difficult than taking any other derivatives, except for an extra step. We need to write the "outside" function in terms of u by figuring out what the "inside" function is.

Example 4.12

Suppose $y = e^{3x}$. Find the derivative of y.

Solution: Here the "inside" function is $3x$ so we write $y = e^u$ where $u = 3x$. Thus the "outside" function is e^u. Notice the variable we use for the "outside" function is now u. We then find the derivative of $y = e^u$ with respect to u, which is simply $y' = e^u$. Next, we find the derivative of u with respect to x, or $u' = 3$. Using the chain rule, we have

$$\frac{dy}{dx} = \frac{dy}{du}\frac{du}{dx}$$
$$= y' \cdot u' \qquad \text{Notice, } y' \text{ is with respect to } u.$$
$$= e^u(3) \qquad \text{Now replace } u.$$
$$= e^{3x}(3)$$
$$= 3e^{3x}.$$

Example 4.13

Suppose $f(x) = (6x^2 + 4x)^3$. Find $f'(x)$.

Solution: Again, we need to find the "inside" function and call it u. Clearly, the "inside" function is $u = 6x^2 + 4x$ and the "outside" function is $y = u^3$. We then find the derivatives of each, $y' = 3u^2$ and $u' = 12x + 4$. Using the chain rule, we have

$$f'(x) = y' \cdot u'$$
$$= 3u^2 \cdot (12x + 4) \qquad \text{Now replace } u.$$
$$= 3(6x^2 + 4x)^2(12x + 4).$$

While we often do some simplification to make the answer look nice, we generally do not fully simplify, particularly in cases like this example. The whole point of using the chain rule in this case is to avoid multiplying polynomials.

Example 4.14

Differentiate the function $f(x) = \sqrt{x^3 - 9}$.

Solution: The "inside" function is $u = x^3 - 9$ and the "outside" function is $y = \sqrt{u} = u^{\frac{1}{2}}$. The derivatives of each are $y' = \frac{1}{2}u^{-\frac{1}{2}} = \frac{1}{2\sqrt{u}}$ and $u' = 3x^2$. Using the chain rule, we have

$$f'(x) = y' \cdot u'$$

$$= \frac{1}{2\sqrt{u}}(3x^2) \qquad \text{Now replace } u.$$

$$= \frac{3x^2}{2\sqrt{x^3 - 9}}.$$

Example 4.15

Differentiate the function $f(x) = 3^{x^2 + x}$.

Solution: The "inside" function is clearly $u = x^2 + x$ and the "outside" function is $y = 3^u$. The derivatives of each of these is $y' = \ln(3)3^u$ and $u' = 2x + 1$. Using the chain rule, we have

$$f'(x) = y' \cdot u'$$

$$= \ln(3)3^u(2x + 1) \qquad \text{Now replace } u.$$

$$= \ln(3)3^{x^2 + x}(2x + 1).$$

4.4 BUSINESS APPLICATIONS OF DIFFERENTIATION

From a business perspective, it does not make sense to try to minimize profits or revenues. In any event, if we wanted zero profits or zero revenues, we would simply choose not to go into business and not to produce anything. However, we are often interested in maximizing profits or revenues. Similarly, in most situations, costs are also minimized when we do not produce anything. But occasionally there are situations where the cost function is unusual and we would like to minimize it. In particular, this sometimes happens because it is cheaper to buy large quantities of materials at a time. (Many manufacturers will give a discount to people who buy a lot of materials at one time. This is sometimes called buying "in bulk.")

In fact, probably the most important application of derivatives in business and economics is to find the quantity that maximizes profit or revenue or minimizes costs. While this is discussed in far greater detail in Chapter 5, these values are found by taking the derivative of the profit, revenue, or cost functions, setting the derivatives equal to zero, and solving for x. However, we will make use of this fact in many of the following examples, though here we are primarily illustrating taking derivatives.

Suppose the cost function for buying a certain item was given by

$$C(x) = \frac{50000}{x} + 0.5x = 50{,}000x^{-1} + 0.5x$$

where x is the number of items we need to purchase. We want to find the number of items we should buy in order to minimize the costs. Below we plot this function for $50 \le x \le 1000$.

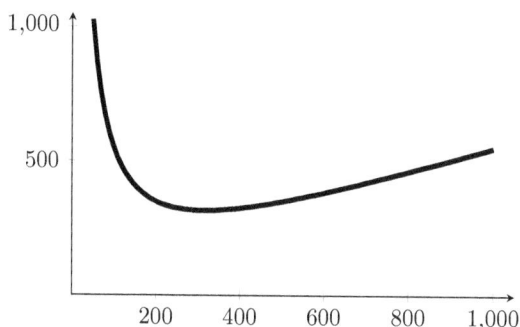

Finding the minimum value requires us to take the derivative, set it equal to zero, and solve for x. Taking the derivative, we have

$$C'(x) = -50{,}000x^{-2} + 0.5$$
$$= \frac{-50{,}000}{x^2} + 0.5.$$

Setting the derivative equal to zero and solving for x gives us

$$\frac{-50{,}000}{x^2} + 0.5 = 0 \implies 0.5 = \frac{50{,}000}{x^2}$$
$$\implies 0.5x^2 = 50{,}000$$
$$\implies x^2 = \frac{50{,}000}{0.5}$$
$$\implies x^2 = 100{,}000$$
$$\implies x = \pm 316.227.$$

Here x represents the number of items so it has to be a whole positive number so the minimum cost occurs when we buy $x = 316$ items. Of course we should technically check to see that his point is a minimum, but that is something we will learn how to do in the next chapter.

Example 4.16

A company buys large quantities of some item which has the cost function $C(x) = \frac{500,000}{x} + 8x$. Find the quantity the company should buy to minimize the cost of the item.

Solution: First we find the derivative of the cost function,

$$C'(x) = -500,000x^{-2} + 8 = \frac{-500,000}{x^2} + 8.$$

Setting the derivative equal to zero and solving for x gives us

$$\frac{-500,000}{x^2} + 8 = 0 \implies x^2 = \frac{500,000}{8} \implies x = \pm 250.$$

Since the quantity of items must be a positive whole number the answer that makes sense in the business context is $x = 250$; this is the number we need to buy to minimize cost. Again, we will skip checking that this point is actually a minimum.

Example 4.17

A company has the following cost function $C(x) = \frac{2x^2+300}{x+5}$. Find the quantity the company should buy to minimize its cost.

Solution: First we find the derivative of the cost function using the quotient rule,

$$C'(x) = \frac{(x+5)(4x) - (2x^2+300)(1)}{(x+5)^2}$$

$$= \frac{2x^2 + 20x - 300}{(x+5)^2}.$$

Setting the derivative equal to zero gives us $2x^2 + 20x - 300 = 0$. This is a quadratic equation. We can solve for x by either factoring, or using the quadratic formula, $x = \frac{-b \pm \sqrt{b^2 - 4ac}}{2a}$, where $a = 2$, $b = 20$, and $c = -300$. Since it does not factor nicely we will use the quadratic equation,

$$x = \frac{-20 \pm \sqrt{20^2 - 4 \cdot 2 \cdot (-300)}}{2 \cdot 2} = \frac{-20 \pm 52.91}{4}.$$

So, $x = 8.23$ or $x = -18.23$. Among the solutions, $x = 8.23$ is the positive solution that makes sense in a production context. Therefore, the company should buy 8 units to minimize the cost.

Example 4.18

Assume the demand $D(p)$ for a product in a competitive market is given by the exponential function $D(p) = 100e^{-0.05p}$, where p is the price per unit.

1. What is the rate of change of demand with respect to price?
2. Identifying revenue maximization and minimization points.

Solution:

1. To find the rate of change of demand with respect to price, we calculate the derivative of the demand function with respect to p:

$$D'(p) = \left(100e^{-0.05p}\right)'$$
$$= 100 \cdot \left(e^{-0.05p}\right)'$$
$$= 100 \cdot (-0.05) \cdot e^{-0.05p}$$
$$= -5e^{-0.05p}.$$

Note that we obtained the derivative using the chain rule directly without the need to manipulate the expression as we did in the last solved example in Chapter 3.

2. Revenue R is given by the product of price and demand, $R(p) = p \cdot D(p)$. To find the maximization and minimization points, we find the derivative of $R(p)$ using the product rule and set it to zero.

$$R(p) = p \cdot 100e^{-0.05p}$$
$$\implies R'(p) = 100e^{-0.05p} + p \cdot (-0.05) \cdot 100e^{-0.05p}$$
$$= 100e^{-0.05p}(1 - 0.05p).$$

Setting $R'(p) = 0$ and solving for p will give the price points for revenue maximization and minimization.

$$R'(p) = 0$$
$$\implies 100e^{-0.05p}(1 - 0.05p) = 0$$
$$\implies 1 - 0.05p = 0 \quad (100e^{-0.05p} \text{ is never zero})$$
$$\implies p = \frac{1}{0.05} = 20.$$

At $p = 20$, the company will either achieve maximum or minimum revenue. The exact nature (maximum or minimum) can be obtained through further analysis as we will see in the coming chapters.

Example 4.19

A company's revenue R and cost C are functions of the quantity q of a product produced and sold. Suppose the revenue function is $R(q) = 200q - q^2$ and the cost function is $C(q) = 500 + 50q$. Find the rate of change of profit with respect to quantity at $q = 50$.

Solution: Profit is given by $P(q) = R(q) - C(q)$.

$$P(q) = R(q) - C(q)$$
$$= (200q - q^2) - (500 + 50q)$$
$$= 150q - q^2 - 500.$$

To find the rate of change of profit with respect to q, we use the power rule:

$$P'(q) = 150 - 2q$$
$$P'(50) = 150 - 2 \cdot 50 = 50.$$

Therefore, the rate of change of profit with respect to the quantity produced and sold at $q = 50$ is \$50. In other words, we can expect our profit to increase by about \$50 when we sell the 51st item.

Example 4.20

Consider the following revenue and cost functions. $R(q) = 200e^{0.05q}$ and $C(q) = 500 + 50q \ln(q)$. Find the rate of change of profit with respect to quantity at $q = 100$.

Solution: First, let us find the profit function $P(q)$:

$$P(q) = R(q) - C(q)$$
$$= 200e^{0.05q} - (500 + 50q \ln(q))$$
$$= 200e^{0.05q} - 50q \ln(q) - 500.$$

Then we differentiate $P(q)$ with respect to q:

$$P'(q) = \frac{d}{dq}(200e^{0.05q}) - \frac{d}{dq}(50q \ln(q)) - \frac{d}{dq}(500)$$
$$= 200(0.05)e^{0.05q} - \left[50 \ln(q) + 50q \cdot \frac{1}{q}\right]$$
$$= 10e^{0.05q} - 50 \ln(q) - 50.$$

Now, calculate $P'(100)$ to find the rate of change of profit at $q = 100$:

$$P'(100) = 10e^{0.05 \cdot 100} - 50\ln(100) - 50$$
$$= 10e^5 - 50\ln(100) - 50$$
$$\approx 1203.87.$$

Thus, the rate of change of profit with respect to the quantity produced and sold at $q = 100$ is 1203.87. In other words, when we sell the 101st item we can expect the profit to increase by about \$1203.87.

Example 4.21

A company's revenue $R(x)$ from selling x units of a product is modeled by the function $R(x) = 500e^{-0.0001x} \cdot x$. Find the optimal number of units to maximize revenue.

Solution: The optimal number of units to maximize revenue can be found by setting the derivative $R'(x)$ equal to zero and then solving for x. Applying the product rule to $R(x)$ we get

$$R'(x) = \frac{d}{dx}\left(500e^{-0.0001x}\right) \cdot x + 500e^{-0.0001x} \cdot \frac{d}{dx}(x)$$
$$= 500\frac{d}{dx}\left(e^{-0.0001x}\right) \cdot x + 500e^{-0.0001x} \cdot 1$$
$$= 500(-0.0001)e^{-0.0001x} \cdot x + 500e^{-0.0001x}$$
$$= 500e^{-0.0001x}(1 - 0.0001x).$$

To find the optimal number of units to maximize revenue, set $R'(x) = 0$ and note that $e^{-0.0001x}$ is never zero. This gives us

$$500e^{-0.0001x}(1 - 0.0001x) = 0$$
$$\Longrightarrow 1 - 0.0001x = 0$$
$$\Longrightarrow x = \frac{1}{0.0001} = 10{,}000.$$

Therefore, the optimal number of units to maximize revenue is 10,000 units.

The final example is a somewhat more advanced than any we have done so far. It involves a composition of three functions, not two, though the principle is the same. For the composition of three or more functions, it is easiest to use Leibniz notation.

Example 4.22

Consider a company whose demand D for a product depends on the marketing expenditure M, which in turn varies with time t, in months. Suppose the demand is given by $D(M(t)) = \sqrt{500 + 20M(t)}$ and the marketing expenditure over time is $M(t) = 100t^2$. Find the rate of change of demand with respect to time when $t = 3$ months.

Solution: We can use the chain rule to find $\frac{dD}{dt}$. Let $u = 500 + 20M(t)$, then $D(u) = \sqrt{u}$. Notice we end up with the composition of three functions; $D\Big(u\big(M(t)\big)\Big)$. This is one instance where the power of Leibniz notation becomes evident.

$$\frac{dD}{dt} = \frac{dD}{du} \cdot \frac{du}{dM} \cdot \frac{dM}{dt}$$

where

$$D(u) = \sqrt{u} \qquad \Longrightarrow \qquad \frac{dD}{du} = \frac{1}{2\sqrt{u}} = \frac{1}{2\sqrt{500 + 20M(t)}}$$

$$u(M) = 500 + 20M \qquad \Longrightarrow \qquad \frac{du}{dM} = 20$$

$$M(t) = 100t^2 \qquad \Longrightarrow \qquad \frac{dM}{dt} = 200t.$$

Therefore,

$$\frac{dD}{dt} = \frac{1}{2\sqrt{500 + 20(100t^2)}} \cdot 20 \cdot 200t.$$

At $t = 3$

$$\frac{dD}{dt} = \frac{20 \cdot 200 \cdot 3}{2\sqrt{500 + 20(100 \cdot 3^2)}} \approx 44.11.$$

Therefore, the rate of change of demand with respect to time at 3 months is approximately 44 units per month.

4.5 PROBLEMS

Question 4.1 *Find the derivatives of the following functions:*

(a) $f(x) = \frac{x^3+5}{x^2+6x}$

(b) $f(x) = \frac{-x^2-x}{x^3-3}$

(c) $f(x) = \frac{e^x}{x^4-x^3}$

(d) $f(x) = \frac{x^5}{x-1}$

(e) $f(x) = \frac{\ln(x)}{x^2-2x}$

(f) $f(x) = \frac{2^x}{\ln(x)}$

(g) $f(x) = \frac{5^x+x}{x^2-e^x}$

(h) $f(x) = \frac{3x^5-2x^3}{-2x^3+5x}$

(i) $f(x) = \frac{\log_5(x)}{2^x}$

Question 4.2 *Find the derivatives of the following functions:*

(a) $f(x) = (x + 7)\ln(x)$

(b) $f(x) = \ln(x)3^x$

(c) $f(x) = (4^x)(3x - 5)$

(d) $f(x) = 2x^3 e^x$

(e) $f(x) = 5x^6 \ln(x)$

(f) $f(x) = (2x^5 + 3x^4)\ln(x)$

(g) $f(x) = 10x(5x^3 - 4x)$

(h) $f(x) = (5x^2 - 3x + 7)e^x$

(i) $f(x) = 5^x \ln(x)$

(j) $f(x) = e^x 7^x$

(k) $f(x) = (x^3 - 2x^2)2^x$

(l) $f(x) = (5^x)(7^x)$

Question 4.3 *Find the derivatives of the following functions:*

(a) $f(x) = 5^{7x}$

(b) $f(x) = -5e^{3x}$

(c) $f(x) = \frac{1}{4}e^{4x}$

(d) $f(x) = \log_7(5x)$

(e) $f(x) = 7^{x^4}$

(f) $f(x) = \ln(2x - 7)$

(g) $f(x) = e^{-3x+5}$

(h) $f(x) = \ln(x^3 - 6x^2)$

(i) $f(x) = (3x - 5)^5$

(j) $f(x) = (2x^2 + 6x)^3$

(k) $f(x) = (5x^3 - 7x^2)^2$

(l) $f(x) = (-4x^4 + 7x^3 - 6x^2)^3$

Question 4.4 *Find the derivatives of the following functions:*

(a) $f(x) = \log_4(3x^2 - 5x + 2)$

(b) $g(x) = \log(x^3 - 5e^x + 2\ln x)$

(c) $g(x) = x^3 - \ln(x^2 + 3x - 1)$

(d) $f(x) = \ln(\frac{3}{x^3})$

(e) $f(x) = (3\ln(x^2) - x + 2)^2$

(f) $g(x) = \frac{e^x + 3x - 1}{e^x - 1}$

Question 4.5 *Find first and second derivatives of the following functions:*

(a) $f(x) = (\log(x^2))^2$

(b) $g(x) = \frac{1}{e^x - 1}$

(c) $g(x) = \log_3(x^2)$

Question 4.6 *A company's profit $P(x)$ is modeled by the function $P(x) = \frac{100x - x^2}{50 + 10x}$, where x represents the number of units sold. Find the rate of change of the profit when $x = 10$ units are sold.*

Question 4.7 *The efficiency $E(v)$ of a wind turbine is given by $E(v) = \frac{v^2}{v^3 + 100}$, where v is the wind speed. Find the rate of change of efficiency when the wind speed is $v = 5$ m/s.*

Question 4.8 *A chemical reaction rate $R(T)$ depends on temperature as $R(T) = \frac{T^2}{T + 100}$, where T is the temperature in degrees Celsius. Determine the rate of change of the reaction rate at a temperature of $T = 10\,°C$.*

Question 4.9 *The concentration $C(x)$ of a medication in the bloodstream is modeled by $C(x) = \frac{200x}{x^2+50}$, where x is the time in hours since the medication was taken. Find the rate of change of concentration when $x = 4$ hours.*

Question 4.10 *A sales model $S(p)$ predicts revenue based on pricing, given by $S(p) = \frac{500p}{p^2+20}$, where p is the price per unit. Calculate the rate of change of revenue when the price per unit is $p = 4$.*

Question 4.11 *A manufacturing company has a cost function for a raw material described by $C(y) = \frac{75000}{y} + y$, where y is the quantity of raw material in kilograms. Determine the optimal quantity of material to minimize costs.*

Question 4.12 *A retail company purchases goods in bulk with the cost function $C(x) = \frac{25000}{x} + 12x$. Find the quantity that minimizes the total cost.*

Question 4.13 *A service provider has a cost function for operations given by $C(x) = \frac{4x^2+150}{x-3}$. Identify the quantity of service units to operate at to minimize costs.*

Question 4.14 *The revenue $R(x)$ from selling x units of a new technology product is modeled by $R(x) = 300xe^{-0.0002x}$. Determine the optimal number of units to sell for maximum revenue.*

Question 4.15 *Consider a product whose market demand $D(p)$ is defined by $D(p) = 80e^{-0.03p}$, with p representing the price per unit. Find the rate of change of demand with respect to price and locate points for revenue maximization and minimization.*

Question 4.16 *A company's profit scenario is modeled with revenue $R(q) = 250q - q^2$ and cost $C(q) = 600 + 30q$. Analyze the rate of change of profit with respect to the quantity produced and sold at $q = 75$.*

Question 4.17 *For a business, the revenue function is $R(q) = 150e^{0.03q}$ and the cost function is $C(q) = 400 + 25q\ln(q)$. Determine the rate of change of profit with respect to the quantity of goods at $q = 150$.*

Question 4.18 *A company's product demand D relies on its advertising expenditure $A(t)$, where $D(A(t)) = \sqrt{1000 + A(t)}$ and $A(t) = 50t^2$. Calculate the rate of change of demand with respect to time at $t = 2$ months.*

CHAPTER 5

Applications of Optimization

In this chapter, we will present several different applications of derivatives. Derivatives can help studying the growth of a function, its minimum and maximum points, its inflection points, and its concavity.

5.1 STATIONARY POINTS

Much important information about a function can be found from its graph. The graph of a function constitutes of all points (x, y) of the Cartesian plan that satisfy $y = f(x)$. A function f is said to be increasing if its graph moves upward from left to right. Mathematically, f is **increasing** if $x_1 < x_2 \implies f(x_1) < f(x_2)$. Similarly, a function f is said to be **decreasing** if its graph moves downward from left to right. Mathematically, f is decreasing if $x_1 < x_2 \implies f(x_1) > f(x_2)$.

Example 5.1

The function $f(x) = 3x^3 + x$ is an increasing function. If you look closely, you can see that the function is always increasing.

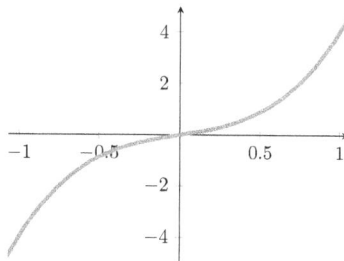

DOI: 10.1201/9781003480235-5

The function $g(x) = -3x^3 - x$ is a decreasing function. Again, if you look closely you can see that it is always decreasing.

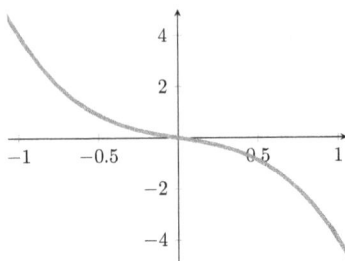

Example 5.2

The graph of the function $f(x) = 2x^3 - 18x^2 + 48x - 27$ in this example is increasing on the intervals $(-\infty, 2)$ and $(4, \infty)$, and decreasing on the interval $(2, 4)$. It is said to be neither increasing nor decreasing when $x = 2$ or when $x = 4$.

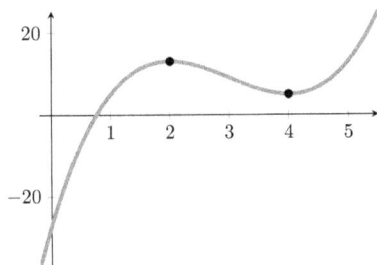

Suppose we were given the function $f(x) = 2x^3 - 18x^2 + 48x - 27$ and were asked if there were a **local maximum** or a **local minimum** to this function. It is clear when we graph the function that there is a local maximum when $x = 2$ and a local minimum when $x = 4$.

Here we are using the word **local** to refer to x values that are close by. The local maximum is shown as the dot on the graph at $x = 2$ and the local minimum is shown as the dot on the graph at $x = 4$. But if we look at the whole graph we can see to the right that the graph goes far higher than the local maximum. And on the left the graph goes far lower than the local minimum. But the point at $x = 2$ is a maximum for x values close to $x = 2$, so we use the phrase local maximum. Similarly, the point at $x = 4$ is a minimum for x values close to $x = 4$, so again we use the phrase local minimum.

But how would we find these without graphing the function? Figuring out if a function is increasing or decreasing and how to find its local maximums and minimums without needing to graph it requires us to notice a few simple things. If we draw the tangent line to the graph at the local maximum, it is clear that the slope of the tangent line is zero. (Below left.) Similarly, if we draw the tangent line to the graph at the local minimum, it is clear that the slope of the tangent line is also equal to zero. (Below right.)

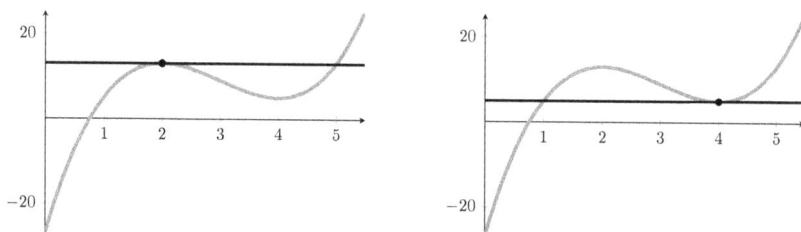

Thus, finding the x-values of the local maximum and minimum points is the same as finding the x-values when the slopes of the tangent lines are equal to zero. And we already know that the slopes of the tangent lines to the graph of a function $f(x)$ are given by the derivative function $f'(x)$. Thus, we need to find out when the derivative function is zero. In other words, we need to set the derivative function equal to zero and then solve for x. First we find the derivative of function $f(x) = 2x^3 - 18x^2 + 48x - 27$ which is

$$f'(x) = 6x^2 - 36x + 48.$$

We then set this equal to zero

$$6x^2 - 36x + 48 = 0.$$

In order to solve for x we simply factor the left-hand side,

$$6x^2 - 36x + 48 = 6(x^2 - 6x + 8) = 6(x - 2)(x - 4).$$

Substituting this back into our equation gives us

$$6(x - 2)(x - 4) = 0,$$

which is only true if

$$x - 2 = 0 \qquad \text{or} \qquad x - 4 = 0,$$

which gives us $x = 2$ or $x = 4$. Now we need to figure out if the function has a local maximum or minimum at $x = 2$ and if the function has a local maximum or minimum at $x = 4$. Actually, there is another possibility, a point of inflection, which will be discussed below. The x values for which the derivative of a function is zero are called **stationary points**. Local maximums and local minimums are two examples of stationary points. They are called "stationary points" because at those points the function $f(x)$ is neither increasing nor decreasing; it is "stationary."

Let us consider three test points separately, one test point to the left of $x = 2$, one test point between $x = 2$ and $x = 4$, and one test point to the right of $x = 4$. It does not matter which points we choose as test points. We have chosen $x = 1$ as the test point to the left of $x = 2$. We can easily determine if the graph of $f(x)$ is increasing or decreasing at $x = 1$ by evaluating the derivative function at $x = 1$,

$$f'(1) = 6(1 - 2)(1 - 4) = 6(-1)(-3) = 18,$$

which is clearly positive. In fact, this positive number is exactly the slope of the tangent line to $f(x)$ at $x = 1$ and it tells us the tangent line is increasing and hence so is the graph of $f(x)$.

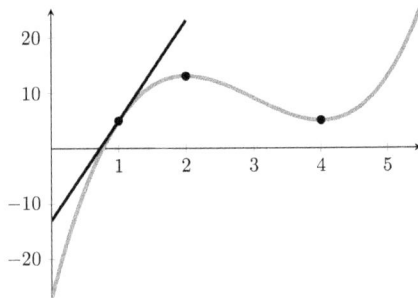

Next, we will choose the test point $x = 3$ which lines between $x = 2$ and $x = 4$. We can easily determine if the graph of $f(x)$ is increasing or decreasing at $x = 3$ by evaluating the derivative function at that point,

$$f'(3) = 6(3 - 2)(3 - 4) = 6(1)(-1) = -6.$$

Since the slope is negative we know the tangent line is decreasing and hence so is the graph. This is shown below. Thus, since the graph of $f(x)$ is increasing to the left of $x = 2$ and decreasing to the right of $x = 2$ we can conclude that the graph has a local maximum at the point $x = 2$. Sometimes you may be asked what point the local maximum is. The x value of the local maximum is 2. To find the y value of the local maximum we evaluate the original function at $x = 2$ to get $f(2) = 13$. Thus the local maximum is at point $(2, 13)$. But usually we are a little more imprecise and just say the maximum happens at $x = 2$.

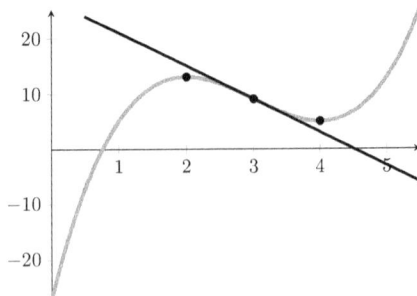

Finally we choose the test point $x = 5$ which lies to the right of $x = 4$. We can easily determine if the graph of $f(x)$ is increasing or decreasing at $x = 5$ by evaluating the derivative function at that point,

$$f'(5) = 6(5-2)(5-4) = 6(3)(1) = 18.$$

Since the slope is positive we know the tangent line is increasing at $x = 5$ and so it the graph of the function. This is shown below. Since the function is decreasing at $x = 3$ and increasing at $x = 5$ we can conclude that the graph has a local minimum at $x = 4$. What point is this local minimum at? We find the y value by evaluating the original function at $x = 4$, $f(4) = 5$, so the local minimum occurs at $(4, 5)$.

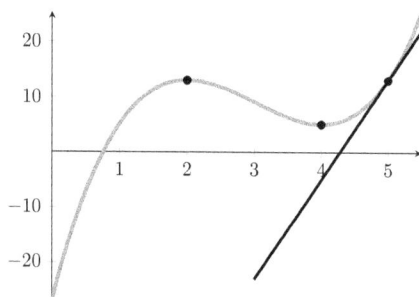

Notice, in the above we used the following facts, $f(x)$ has a stationary point when $f'(x) = 0$, that if $f(x)$ is increasing then $f'(x) > 0$, and if $f(x)$ is decreasing then $f'(x) < 0$. We summarize this below:

$f(x)$	$f'(x)$
stationary	zero
increasing	positive
decreasing	negative

Now consider the function $f(x) = x^4 - 8x^3 + 18x^2 - 12$ graphed below. This function is a little different from the last one. It clearly has a local minimum at $x = 0$ but something interesting also happens at $x = 3$. The graph of the function is increasing between $x = 0$ and $x = 3$ but then at $x = 3$ the graph flattens out, only to start increasing again after $x = 3$. This point is an example of an **inflection point**. We will discuss inflection points in greater detail soon, but for now, notice the slope of the tangent line to the graph at $x = 3$ is horizontal too, just like the tangent lines at local maximums and minimums.

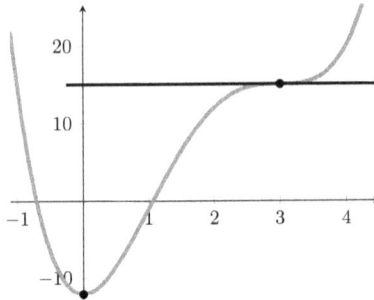

Like before we can take the derivative of $f(x)$, set it equal to zero, and solve for x. First we find the derivative of the function $f(x) = x^4 - 8x^3 + 18x^2 - 12$,

$$f'(x) = 4x^3 - 24x^2 + 36x.$$

We then set this equal to zero

$$4x^3 - 24x^2 + 36x = 0.$$

In order to solve for x we simply factor the left-hand side,

$$4x^3 - 24x^2 + 36x = 4x(x^2 - 6x + 9) = 4x(x - 3)(x - 3).$$

Substituting back into our equation gives us

$$4x(x - 3)(x - 3) = 0,$$

which is only true if

$$x = 0 \qquad \text{or} \qquad x - 3 = 0,$$

which gives us $x = 0$ or $x = 3$. Determining the nature of the points at $x = 0$ and $x = 3$ proceeds just like before, we have to choose test points to the left of $x = 0$, between $x = 0$ and $x = 3$, and to the right of $x = 3$. We then find the slopes of the tangent lines at those test points using the derivative function and use this to determine if the graph is increasing or decreasing.

$x = -1$	$x = 2$	$x = 4$
$f'(-1) = -64$	$f'(2) = 8$	$f'(4) = 16$
negaive slope	positive slope	positive slope
$\Rightarrow$ decreasing	$\Rightarrow$ increasing	$\Rightarrow$ increasing

Since the graph is decreasing before $x = 0$ and increasing after $x = 0$ then clearly $x = 0$ is a minimum. But notice that the graph is increasing between $x = 0$ and $x = 3$ and also increasing after $x = 3$. This is what happens at a point of inflection. It is also possible that the graph could be decreasing both before and after a point of inflection. How you tell a stationary point is also a point of inflection is that the graph is "going in the same direction" both before and after the point. That "direction" may either be increasing or decreasing. Below we draw the tangent lines to these three points on the graph at $x = -1$, $x = 2$, and $x = 4$.

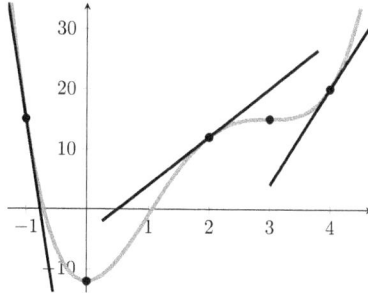

The phrase **stationary points** refers to points at which the tangent line has zero slope. If something is stationary it does not move. The tangent lines at stationary points neither increase nor decrease and so in a sense "do not move" and are "stationary." Stationary points can be categorized as either local maximums, local minimums, or inflection points. Putting everything discussed above together, we have a procedure to find stationary points and to determine if they are local maximums, local minimums, or points of inflection.

First Derivative Test: Procedure to find stationary points of a function $f(x)$ and determine their nature (local maximum, local minimum, or point of inflection.)

1. Find $f'(x)$.
2. Set $f'(x) = 0$.
3. Solve for x. These are the x values of the stationary points.
4. Determine the nature of each stationary point x_{sp} by choosing a test point x_{test} before and after each stationary point and using the below table. The below table is usually called the **first derivative test**.

$x_{test} < x_{sp}$	x_{sp}	$x_{sp} < x_{test}$	Nature of Point
$f'(x_{test}) > 0$	$f'(x_{sp}) = 0$	$f'(x_{test}) < 0$	$\Rightarrow$ local maximum
$f'(x_{test}) < 0$	$f'(x_{sp}) = 0$	$f'(x_{test}) > 0$	$\Rightarrow$ local minimum
$f'(x_{test}) > 0$	$f'(x_{sp}) = 0$	$f'(x_{test}) > 0$	$\Rightarrow$ point of inflection
$f'(x_{test}) < 0$	$f'(x_{sp}) = 0$	$f'(x_{test}) < 0$	$\Rightarrow$ point of inflection

Note: To find the y values of the stationary points substitute the x values into $f(x)$.

In many business and economics applications, we are specifically interested in the behavior of a function in some specific closed interval. For example, if the variable x represents the number of items that could be produced in a factory that can produce a maximum of 1000 items a day, our range of interest may be $[0, 1000]$. In our applications we may not only be interested in local maximums and minimums, but we may want to know the **absolute maximum** or **absolute minimum** on our range. According to the **Extreme Value Theorem**, which we will not prove, the only way to guarantee an absolute maximum and minimum exist is if our function is continuous and restricted to a closed interval $[a, b]$. Absolute maxima/minima are also called **global maxima/minima**.

Extreme Value Theorem:

Given a continuous function $f(x)$ on a closed interval $[a, b]$, then the **absolute maximum** and **absolute minimum** of $f(x)$ exist and are located either at a stationary point x_{sp} in the interval $[a, b]$ or at one of the interval endpoints $x = a$ or $x = b$.

To find the absolute, or global, maximum or minimum values we only need to find the y values of the candidate points. The point with the largest y value is the absolute maximum and the point with the smallest y value is the absolute minimum.

Example 5.3

Find the local and absolute extrema for $f(x) = \frac{1}{4}x^4 - x^3 + x^2 + 1$ on the closed interval $[0, 3]$.

Solution: First we find the derivative of the function,

$$f'(x) = x^3 - 3x^2 + 2x.$$

We then set this equal to zero and solve,

$$x^3 - 3x^2 + 2x = 0 \implies x(x^2 - 3x + 2) = 0$$
$$\implies x(x - 1)(x - 2) = 0$$
$$\implies x = 0 \text{ or } x - 1 = 0 \text{ or } x - 2 = 0$$
$$\implies x = 0 \text{ or } x = 1 \text{ or } x = 2.$$

Second, we determine the nature of the points $x = 0$, $x = 1$, and $x = 1$, by choosing four test values. Note, the actual test values do not matter, you may choose different test values, as long as the values are in the appropriate range.

$x = -1$	$x = 0.5$	$x = 1.5$	$x = 3$
$f'(-1) = -6$	$f'(0.5) = \dfrac{3}{8}$	$f'(1.5) = -\dfrac{3}{8}$	$f'(3) = 6$
negative slope	positive slope	negative slope	positive slope
$\Rightarrow$ decreasing	$\Rightarrow$ increasing	$\Rightarrow$ decreasing	$\Rightarrow$ increasing

Therefore, there are local minimums at $x = 0$ and at $x = 2$. There is also a local maximum at $x = 1$. Now we are interested in finding the absolute extrema on the closed interval $[0, 3]$. Thus we need to also consider the endpoints $x = 0$ and $x = 3$. To find the absolute, or global, maximum and minimum points we need to consider the y values of the candidate points. In this case $x = 0$ is already a local minimum, but in general that will not be the case. Considering both the stationary points and the interval's endpoints, we have $f(0) = 1, f(1) = \frac{5}{4}, f(2) = 1, f(3) = \frac{13}{4}$.

We deduce that $(0, 1)$ is a local minimum, $(2, 1)$ is another local minimum, $(1, \frac{5}{4})$ is a local maximum, and there is a maximum at $(3, \frac{13}{4})$. Since $f(0) = f(2)$ they are both absolute minimum, while $(3, \frac{13}{4})$ is the absolute maximum. This is obvious when we look at the graph below.

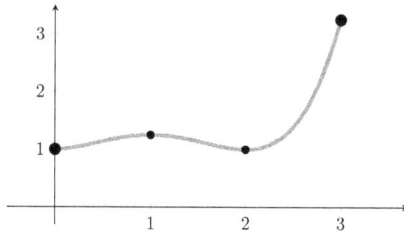

Example 5.4

The number of people visiting a bookshop since its opening in 2001 is giving by the function

$$f(x) = 2x^3 - 45x^2 + 300x + 500.$$

Find the local extrema of the function f from 2001 to 2021.

Solution: Since the bookshop opened in 2001 we consider $x = 0$ as the starting point, which corresponds to the year 2001, and $x = 20$ as the endpoint, which corresponds to the year 2021. Thus we are considering the closed interval $[0, 20]$. First we find the derivative,

$$f'(x) = 6x^2 - 90x^2 + 300.$$

We then set this equal to zero and solve

$$
\begin{aligned}
6x^2 - 90x + 300 &\implies 6(x^2 - 15x + 50) = 0 \\
&\implies 6(x - 5)(x - 10) = 0 \\
&\implies x - 5 = 0 \text{ or } x - 10 = 0 \\
&\implies x = 5 \text{ or } x = 10.
\end{aligned}
$$

Now determine the nature of $x = 5$ and $x = 10$ using test values:

$x = 0$	$x = 6$	$x = 20$
$f'(0) = 300$	$f'(6) = -24$	$f'(20) = 900$
positive slope	negative slope	positive slope
$\Rightarrow$ increasing	$\Rightarrow$ decreasing	$\Rightarrow$ increasing

Therefore there is a local maximum at $x = 5$ and a local minimum at $x = 10$. As we are looking for extrema on $[0, 20]$, we will need to also consider the two endpoints of the closed interval; $f(0)$ and $f(20)$. Since

$$f(0) = 500, \ f(5) = 1125, \ f(10) = 1000, \ f(20) = 4500,$$

we deduce that $(20, 4500)$ is the absolute maximum and $(0, 500)$ is the absolute minimum

Example 5.5

Find the local extrema for $f(x) = x^3$.

Solution: Notice that we are not given a closed interval so the Extreme Value Theorem does not apply. First we find the derivative of $f'(x) = 3x^2$ and then set this equal to zero and solve for x,

$$3x^2 = 0 \implies x = 0.$$

Second we determine the nature of the point $x = 0$ by choosing two test values, one less than 0 and one greater than 0.

$x = -1$	$x = 1$
$f'(-1) = 3$	$f'(1) = 3$
positive slope	positive slope
$\Rightarrow$ increasing	$\Rightarrow$ increasing

Therefore, the function $f(x) = x^3$ does not have a local minimum nor a local maximum at $x = 0$, it has an inflection point. Indeed, we can see this from looking at the graph of the function.

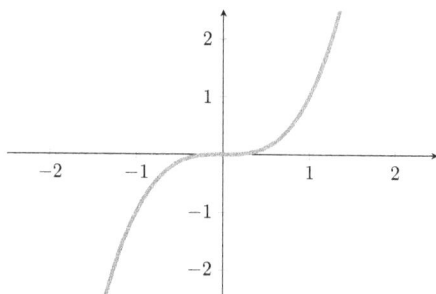

Example 5.6

Given $f(x) = 10 - 12x + 6x^2 - x^3$. Find the intervals on which $f(x)$ is increasing and the intervals on which $f(x)$ is decreasing.

Solution: First we find the derivative $f'(x) = -12 + 12x - 3x^2$. We then set this equal to zero and solve it

$$-12 + 12x - 3x^2 = 0 \implies -3(4 - 4x + x^2) = 0$$
$$\implies -3(2 - x)^2 = 0$$
$$\implies 2 - x = 0$$
$$\implies x = 2.$$

Second we determine the nature of the point $x = 2$ by choosing two values one less than $x = 2$ and one greater than $x = 2$.

$x = 0$	$x = 3$
$f'(0) = -12$	$f'(3) = -1$
negative slope	negative slope
$\Rightarrow$ decreasing	$\Rightarrow$ decreasing

Therefore, the function $f(x) = 1 - 12x + 6x^2 - x^3$ does not have a local minimum or a local maximum at $x = 2$ and it is decreasing on $(-\infty, 2) \cup (2, \infty)$. Since $f'(2) = 0$ we say that the function is neither increasing nor decreasing at $x = 2$.

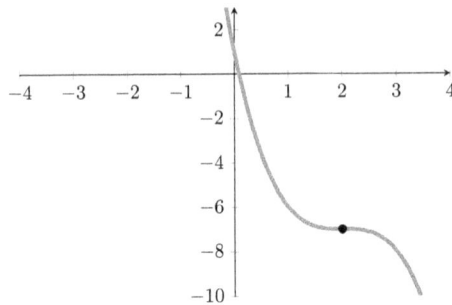

Example 5.7

Given $f(x) = x^3 + 3x^2 - 24x$. Find the intervals on which $f(x)$ is increasing and the intervals on which $f(x)$ is decreasing.

Solution: First we find the derivative $f'(x) = 3x^2 + 6x - 24$. We then set this equal to zero and solve it

$$3x^2 + 6x - 24 = 0 \implies 3(x^2 + 2x - 8) = 0$$
$$\implies 3(x - 2)(x + 4) = 0$$
$$\implies x - 2 = 0 \text{ or } x + 4 = 0$$
$$\implies x = 2 \text{ or } x = -4.$$

Second we determine the nature of the points $x = -4$ and $x = 2$, by choosing three test values.

$x = -5$	$x = 0$	$x = 3$
$f'(-5) = 21$	$f'(0) = -24$	$f'(3) = 21$
positive slope	negative slope	positive slope
$\Rightarrow$ increasing	$\Rightarrow$ decreasing	$\Rightarrow$ increasing

Therefore there is a local minimum at $x = 2$ and a local maximum at $x = -4$. The function $f(x)$ is increasing on $(-\infty, -4) \cup (2, \infty)$ and decreasing on $(-4, 2)$.

5.2 HIGHER-ORDER DERIVATIVES

Before moving on to the next section we need to introduce the second derivative. Once you know about derivatives, second derivatives are easy to understand. A second derivative is simply the derivative of the derivative. Consider the function we had looked at in the last section,

$$f(x) = 2x^3 - 18x^2 + 48x - 27.$$

We had found the derivative of this function to be

$$f'(x) = 6x^2 - 36x + 48.$$

This is often called the **first derivative of** $f(x)$. But notice, the first derivative of $f(x)$, usually written as $f'(x)$ or as $\frac{df}{dx}$ is also a function of x. Since $f'(x)$ is a function of x we can also take the derivative of this function. The derivative of the first derivative function $f'(x)$ is called the **second derivative of** $f(x)$. Thus, the second derivative of this function would be

$$f''(x) = 12x - 36.$$

Notice that there are two little dashes, called primes, above the f. The fact that there are two primes tells you that this is a second derivative of the function f. In Leibniz notation this would be written as

$$\frac{d^2 f}{dx^2} = 12x - 36.$$

The exponent two clearly indicates that this is the second derivative. but the placement of the twos perhaps needs a little explanation. Consider the first derivative of $f(x)$ in Leibniz notation. We could write it this way;

$$\frac{df}{dx} = \frac{d}{dx}(f).$$

Here we could interpret the $\frac{d}{dx}$ as meaning "take the derivative" of the function f with respect to the variable x. Thus, since the second derivative is taking the derivative of $\frac{d}{dx}(f)$, which we could write as

$$\frac{d}{dx}\left(\frac{d}{dx}(f)\right) = \frac{d^2}{dx^2}(f) = \frac{d^2 f}{dx^2}.$$

Of course, there is no reason why we need to stop at the second derivative. We can take **third derivatives**, **fourth derivatives**, and so on. For third derivatives we often use three primes, so the third derivative of the function $f(x)$ would be

$$f'''(x) = \frac{d^3 f}{dx^3} = 12.$$

Counting primes in our Newtonian notation starts to get annoying so for derivatives higher than three we often would write a number in parenthesis. For example, the fourth derivative of $f(x)$ would be

$$f^{(4)}(x) = \frac{d^4 f}{dx^4} = 0,$$

and so on. The following table gives the notation.

Newtonian	Leibniz	Meaning
$f'(x)$	$\frac{df}{dx}$	the first derivative of f
$f''(x)$	$\frac{d^2 f}{dx^2}$	the second derivative of f
$f'''(x)$	$\frac{d^3 f}{dx^3}$	the third derivative of f
$f^{(4)}(x)$	$\frac{d^4 f}{dx^4}$	the fourth derivative of f
$\vdots$	$\vdots$	$\vdots$
$f^{(n)}(x)$	$\frac{d^n f}{dx^n}$	the nth derivative of f

Example 5.8

Find the first through fifth derivatives of $f(x) = x^4 - 3x^3 + 2x - 1$.

Solution:
$$f(x) = x^4 - 3x^3 + 2x - 1$$
$$\implies f'(x) = 4x^3 - 9x^2 + 2$$
$$\implies f''(x) = 12x^2 - 18x$$
$$\implies f'''(x) = 24x - 18$$
$$\implies f^{(4)}(x) = 24$$
$$\implies f^{(5)}(x) = 0$$

Example 5.9

Find the first through sixth derivatives of $f(x) = 2^x + e^{3x}$.

Solution:
$$f(x) = 2^x + e^{3x}$$
$$f'(x) = \ln(2) \cdot 2^x + 3e^{3x}$$
$$f''(x) = [\ln(2)]^2 \cdot 2^x + 3^2 e^{3x}$$
$$f'''(x) = [\ln(2)]^3 \cdot 2^x + 3^3 e^{3x}$$
$$f^{(4)}(x) = [\ln(2)]^4 \cdot 2^x + 3^4 e^{3x}$$

For the first term, we use the derivative rule for exponential functions and for the second term, we use the chain rule. At this point, it is fairly easy to see the pattern that has emerged.

While it is important to understand that derivatives can be taken more than once, in this book we will generally have no reason to take derivatives any higher than the second derivative.

5.3 POINTS OF INFLECTION

The second derivative of a function can be used to find about the function's concavity through the inflection points. In this section we will explain in greater detail what **inflection points** are and how they are used to find the function concavity. Consider the graph of the function $f(x) = x^3 - 12x^2 + 45x - 48$ below.

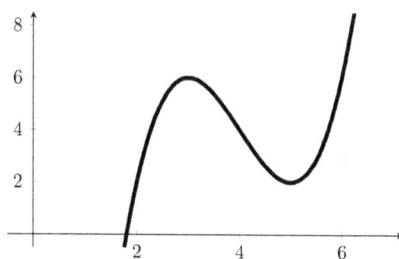

On the left the graph is somehow "facing downwards" while on the right it is somehow "facing upwards." When a graph is "facing downwards" like this one does on the left we call it **concave down**. (Like a frown.) And when a graph is "facing upwards" on the right we call it **concave up**. (Like a cup.) Below, we have graphed the concave down part of $f(x)$ in light gray and the concave up part of $f(x)$ in dark gray. The point where the graph changes from concave down to concave up, shown show as a black dot, is called an **inflection point**. But why do we say that particular point is where concave down turns to concave up?

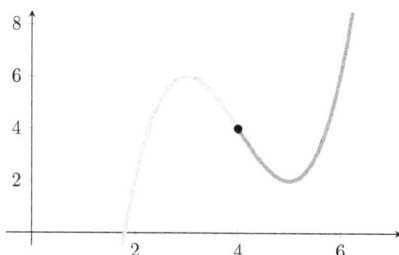

A function $f(x)$ is defined to be concave down as long as the slopes of its tangent lines are decreasing as one moves from left to right. And it is concave up as long as the slopes of its tangent lines are increasing as one moves from left to right. We already know that the derivative function $f'(x)$ gives the slopes of the tangent lines. This means that a function $f(x)$ is concave down when $f'(x)$ is decreasing and concave up when $f'(x)$ is increasing.

Also, from the last section we know that if $f(x)$ is decreasing then $f'(x) < 0$ and if $f(x)$ is increasing then $f'(x) > 0$. In the definition of concave down we had $f'(x)$ is decreasing, which means that $f''(x) < 0$. Similarly, in the definition of concave up we had $f'(x)$ is increasing, which means that $f''(x) > 0$. Summarizing, we have the following:

Definition of Concavity

- A function $f(x)$ is said to be **concave down** on an interval if the slopes of the tangent lines are decreasing over that interval. That is, when the derivative function $f'(x)$ is decreasing. This means that $f''(x) < 0$.
- A function $f(x)$ is said to be **concave up** on an interval if the slopes of the tangent lines are increasing over that interval. That is, when the derivative function $f'(x)$ is increasing. This means that $f''(x) > 0$.

Let us take a closer look at all of this graphically to convince ourselves it is true and to get a better understanding of inflection points. Below the function $f(x) = x^3 - 12x^2 + 45x - 48$ is graphed in black and its derivative function $f'(x) = 3x^2 - 24x + 45$ is graphed in gray. It is easy to see (or check) that $f'(x)$ has a local minimum at $x = 4$. The line $x = 4$ is graphed as a dashed vertical line. To the left of line $x = 4$ the function $f'(x)$ is decreasing. That means the slopes of the tangent lines to $f(x)$ are decreasing, and so $f(x)$ is concave down in this region. To the right of the line $x = 4$ the function $f'(x)$ is increasing. That means the slopes of the tangent lines to $f(x)$ are increasing, and so $f(x)$ is concave up in this region.

But if you think about it, when $f'(x)$ is decreasing means the slopes of the tangent lines of $f'(x)$ are negative, which is the same thing as saying $f''(x)$ is negative. And when $f'(x)$ is increasing means the slopes of the tangent lines to $f'(x)$ are positive, which is the same thing as saying $f''(x)$ is positive. And since the inflection points happen at the local maxima or minima of $f'(x)$ this is equivalent to saying the inflection points happen when $f''(x) = 0$. But caution: **Points of inflection happen when $f''(x) = 0$, but not every value x for which $f''(x) = 0$ is an inflection point.** Inflection points happen at values of x for which $f''(x) = 0$ and there is a change in concavity. We will see an example later where $f''(x) = 0$ but there is no change in concavity at that point.

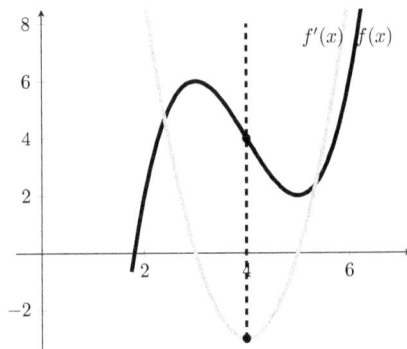

Now let us return to our example from the previous section, $f(x) = x^4 - 8x^3 + 18x^2 - 12$, shown below in black. The derivative $f'(x) = 4x^3 - 24x^2 + 36x$ is shown in gray. Clearly the function $f'(x)$ has both a local maximum and a local minimum. These correspond to two points of inflection on the original function $f(x)$. Only the second point of inflection, the one at $x = 3$, is also a stationary point of $f(x)$. The first point of inflection, at $x = 1$, is not a stationary point of $f(x)$. This is an important point: **Only some stationary points are also inflection points and only some inflection points are also stationary points.**

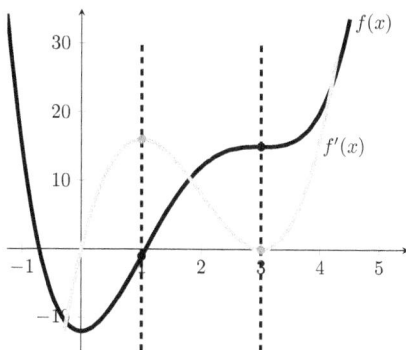

But if you look at the below graph, which shows $f(x) = x^4 - 8x^3 + 18x^2 - 12$ in black and $f''(x) = 12x^2 - 48x + 36$ in gray, and think about it carefully, when $f'(x)$ is decreasing this means the slopes of the tangent lines of $f'(x)$ are negative, which is the same thing as saying $f''(x)$ is negative. And when $f'(x)$ is increasing this means the slopes of the tangent lines to $f'(x)$ are positive, which is the same thing as saying $f''(x)$ is positive. And since the inflection points happen at the local maxima or minima of $f'(x)$ this is equivalent to saying the inflection points happen when $f''(x) = 0$. Thus we can see that when $f''(x) < 0$, we have $f(x)$ is concave down, and when $f''(x) > 0$, we have $f(x)$ is concave up.

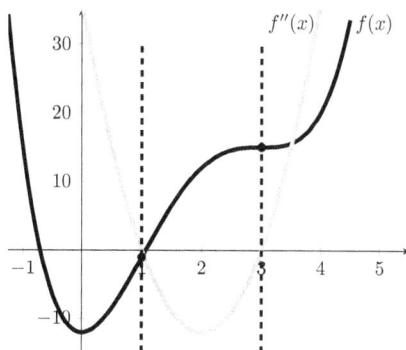

Procedure for finding points of inflection for a function $f(x)$

1. Find $f''(x)$.
2. Set $f''(x) = 0$ and solve for x. These are the possible x values of the points of inflection.
3. Choose test points x_{test} less than and greater than each possible point of inflection and use $f''(x_{test})$ to find the concavity of $f(x)$. If there is a concavity change, then x is a point of inflection. If there is no concavity change, then x is not a point of inflection.
4. To find the y values of the points of inflection substitute x into $f(x)$.

From what was explained above about the relationships between first and second derivative we obtain a new procedure for determining whether a stationary point is a local maximum or a local minimum using the second derivative. This procedure is called the **second derivative test**.

The Second Derivative Test for Local Extrema

1. Find $f'(x)$ and $f''(x)$.
2. Solve $f'(x) = 0$ to find the stationary points x_{sp}, i.e. $f'(x_{sp}) = 0$.
3. If $f''(x_{sp}) > 0$, then there is a local minimum at x_{sp}.
4. If $f''(x_{sp}) < 0$, then there is a local maximum at x_{sp}.
5. If $f''(x_{sp}) = 0$ we need to use the first derivative test.

Example 5.10

Given $f(x) = x^7$, find its inflection points.

Solution: We first find the first and second derivatives

$$f'(x) = 7x^6 \quad \text{and} \quad f''(x) = 42x^5.$$

We then solve $f''(x) = 0$ getting

$$42x^5 = 0 \implies x = 0.$$

We next choose test points to the left and right of $x = 0$. We will choose $x = -1$ to the left and $x = 1$ to the right. This gives us

$$f''(-1) = -42 < 0 \implies f(x) \text{ is concave down,}$$
$$f''(1) = 42 > 0 \implies f(x) \text{ is concave up.}$$

Since there is a change of inflection, then $x = 0$ is an inflection point. Since $f(0) = 0$, the point $(0,0)$ is the inflection point of $f(x) = x^7$.

Example 5.11

Given $f(x) = x^6$, find its inflection points.

Solution: We first find the first and second derivatives

$$f'(x) = 6x^5 \quad \text{and} \quad f''(x) = 30x^4.$$

We then solve $f''(x) = 0$,

$$30x^4 = 0 \implies x = 0.$$

We next choose test points to the left and right of $x = 0$. We will choose $x = -1$ to the left and $x = 1$ to the right. This gives us

$$f''(-1) = 30 > 0 \implies f(x) \text{ is concave up,}$$
$$f''(1) = 30 > 0 \implies f(x) \text{ is concave up.}$$

There is no change of inflection so $x = 0$ is not an inflection point. Indeed, $f(x) = x^6$ does not have an inflection point. This is an example where $f''(x) = 0$ does not give an inflection point.

Example 5.12

Given $f(x) = \frac{2}{3}x^3 - 3x^2 + 4x + 1$, study its concavity.

Solution: We first find the first and second derivatives,

$$f'(x) = 2x^2 - 6x + 4 \quad \text{and} \quad f''(x) = 4x - 6.$$

We then solve $f''(x) = 0$,

$$4x - 6 = 0 \implies x = 1.5,$$

which is the possible inflection point for $f(x)$. We find the sign of the second derivative both to the left and right of the point $x = 1.5$ by choosing two test points, say $x = 0$ and $x = 2$.

$x = 0$	$x = 2$
$f''(0) = -6 < 0$	$f''(2) = 2 > 0$
$\Rightarrow$ concave down	$\Rightarrow$ concave up

Therefore, the function $f(x)$ is concave down on $(-\infty, 1.5)$ and concave up on $(1.5, \infty)$, which means there is a concavity change at $x = 1.5$. Since $f(1.5) = 2.5$, the point $(1.5, 2.5)$ is the inflection point.

Example 5.13

Consider again the function $f(x) = \frac{2}{3}x^3 - 3x^2 + 4x + 1$ from example 5.12. Find its local extrema.

Solution: From Example 5.12, we know that $x = 1.5$ is an inflection point of the function $f(x)$. Now find the stationary points by finding the derivative of the function $f(x)$,

$$f'(x) = 2x^2 - 6x + 4.$$

Now we solve $f'(x) = 0$,

$$2x^2 - 6x + 4 = 0 \implies 2(x^2 - 3x + 2) = 0$$
$$\implies 2(x - 1)(x - 2)) = 0$$
$$\implies x - 1 = 0 \text{ or } x - 2 = 0$$
$$\implies x = 1 \text{ or } x = 2.$$

We determine the nature of $x = 1$ and $x = 2$, by choosing test values:

$x = 0$	$x = 1.5$	$x = 3$
$f'(0) = 4 > 0$	$f'(0) = -0.5 < 0$	$f'(3) = 4 > 0$
positive slope	negative slope	positive slope
$\Rightarrow$ increasing	$\Rightarrow$ decreasing	$\Rightarrow$ increasing

Therefore, there is a local minimum at $x = 2$ and a local maximum at $x = 1$. If we instead used the new procedure of the second derivative test, we would substitute the two stationary points into the second derivative $f''(x) = 4x - 6$ as follows:

$x = 1$	$x = 2$
$f''(1) = -2 < 0$	$f''(2) = 2 > 0$
a local maximum at $x = 1$	a local minimum at $x = 2$

Both procedures yielded the same results.

Example 5.14

Find the local extrema of the function $f(x) = 3x^4 - 2x^3 + 1$.

Solution: First, we find the first and second derivatives:

$$f'(x) = 12x^3 - 6x^2 \quad \text{and} \quad f''(x) = 36x^2 - 12x.$$

Next, solve $f'(x) = 0$.

$$12x^3 - 6x^2 = 0 \implies 6x^2(2x - 1) = 0$$
$$\implies x^2 = 0 \text{ or } 2x - 1 = 0$$
$$\implies x = 0 \text{ or } x = \frac{1}{2}.$$

Now we use the second derivative test:

$x = 0$ $\qquad\qquad\qquad\qquad x = \dfrac{1}{2}$

$f''(0) = 0$ $\qquad\qquad\qquad f''\left(\dfrac{1}{2}\right) = 3 > 0$

no decision at $x = 0$ $\qquad\quad$ a local minimum at $x = \dfrac{1}{2}$

Since it was not possible to determine the nature of the stationary point $x = 0$, we must use the first derivative test by choosing two test points, one to the left and one to the right of $x = 0$. The test point to the right of $x = 0$ needs to be between $x = 0$ and $x = \frac{1}{2}$.

$x = -1$ $\qquad\qquad\qquad\qquad x = 0.1$

$f'(-1) = -18 < 0$ $\qquad\qquad f'(0.1) = -0.047 < 0$

negative slope $\qquad\qquad\qquad$ negative slope

$\Rightarrow$ decreasing $\qquad\qquad\qquad \Rightarrow$ decreasing

Thus, $x = 0$ is not a local extrema. We can see this from the graph of the function.

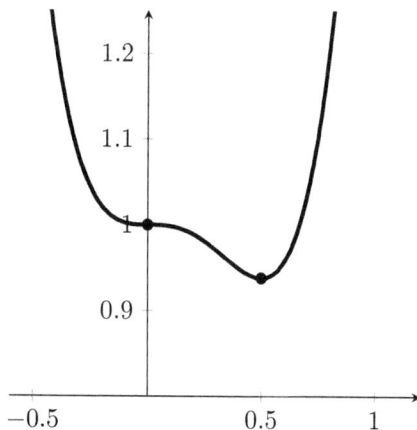

Example 5.15

Recall in Chapter 3, we discussed the logarithmic growth function

$$N(t) = \frac{L}{1 + Ce^{-kt}}.$$

There we said that the point of diminishing returns occurs at the point $t = \frac{\ln(C)}{k}$. The point of diminishing returns is the point when the instantaneous rate of change of the function stops increasing and starts decreasing. But this is exactly when the slopes of the tangent lines stop increasing and start decreasing. In other words, when the function goes from concave up to concave down. We now have the tools necessary to find this inflection point.

Solution: Since

$$N(t) = \frac{L}{1 + Ce^{-kt}} = L(1 + Ce^{-kt})^{-1}$$

we can use the chain rule twice to take the derivative,

$$N'(t) = L(-1)(1 + Ce^{-kt})^{-2}(-kCe^{-kt})$$
$$= kLCe^{-kt}(1 + Ce^{-kt})^{-2}.$$

Taking the second derivative requires both the product rule and multiple applications of the chain rule,

$$N''(t) = kLC\left[-ke^{-kt}(1 + Ce^{-kt})^{-2} + e^{-kt}(-2)(1 + Ce^{-kt})^{-3}(-kCe^{-kt})\right]$$
$$= kLC\left[\frac{-ke^{-kt}}{(1 + Ce^{-kt})^2} - \frac{2e^{-kt}(-kCe^{-kt})}{(1 + Ce^{-kt})^3}\right]$$
$$= kLC\left[\frac{-ke^{-kt}(1 + Ce^{-kt}) + 2kCe^{-2kt}}{(1 + Ce^{-kt})^3}\right]$$
$$= kLC\left[\frac{-ke^{-kt} + kCe^{-2kt}}{(1 + Ce^{-kt})^3}\right].$$

Setting the second derivative equal to zero and solving for t we get

$$N''(t) = 0$$
$$\Longrightarrow kLC\left[\frac{-ke^{-kt} + kCe^{-2kt}}{(1 + Ce^{-kt})^3}\right] = 0$$

$$\implies -ke^{-kt} + kCe^{-2kt} = 0$$
$$\implies kCe^{-2kt} = ke^{-kt}$$
$$\implies e^{-kt} = \frac{1}{C}$$
$$\implies C = e^{kt}$$
$$\implies \ln(C) = kt$$
$$\implies t = \frac{\ln(C)}{k},$$

which is what we wanted to find. Indeed, this was the formula given in Chapter 3. We should, of course, double check that there is a change in concavity at this point, but we will forgo that here. Finally, evaluating the function at this point will give us the y-value.

5.4 OPTIMIZATION IN BUSINESS SCENARIOS

There are many applications of derivatives in economics and business. In Chapter 2, we introduced the concepts of marginal cost, revenue, and profit. In this chapter, we will review those applications before presenting some additional applications of the derivatives, such as finding maximum revenue and profit in a monopoly regime and diminishing marginal returns. We begin by noting some basic relationships that come from business and economics.

Revenue, Cost, and Profit:

If x is the quantity produced, p is price, FC is fixed costs, and VC is variable costs then:

$$R(x) = p \times x, \qquad \text{Revenue Function}$$
$$C(x) = FC + VC, \qquad \text{Total Cost Function}$$
$$P(x) = R(x) - C(x), \qquad \text{Profit Function}$$

Marginal Cost and Marginal Revenue

As we have seen, business people often discuss the **marginal cost** of producing some item. In business, the marginal cost is defined to be the additional cost of producing one additional unit of some product or service. The marginal cost can be closely approximated using the derivative of the cost function. Thus, marginal cost is generally considered a synonym of derivative. Let us see how this works in an example.

Suppose a company was producing widgets and the cost function for producing x widgets a week was given by

$$C(x) = 6000 + 240x - 0.05x^2.$$

A business person may want to know what the marginal cost of producing 300 widgets is. They would calculate this by first taking the derivative of the cost function,

$$C'(x) = 240 - 0.1x,$$

which is often called the **marginal cost function**. To find the marginal cost of producing 300 widgets they would then substitute 300 into the derivative function,

$$C'(300) = 240 - 0.1(300) = 210.$$

Thus the marginal cost of making 300 widgets per week is 210 dollars (or whatever unit the money is in).

But what does this actually mean? Using the definition for marginal cost given above, the marginal cost is the cost added by producing one additional unit of some product or service. So, when the company is producing 300 widgets a week it would cost them 210 dollars more to increase production from 300 widgets a week to 301 widgets a week. Another way of saying this is that if production is 300 widgets per week, the costs are increasing at a rate of 210 dollars per widget.

As was mentioned above, the marginal cost is a good approximation. Let us see how good this approximation is. To find the exact cost of making the 301st widget, we would need to calculate the exact cost of making 301 widgets and then subtract off the exact cost of making 300 widgets,

$$C(301) = 6000 + 240(301) - 0.05(301)^2 = 73{,}709.95,$$
$$C(300) = 6000 + 240(300) - 0.05(300)^2 = 73{,}500.00,$$

which gives us the exact cost of making the 301st widget to be

$$C(301) - C(300) = 73{,}709.95 - 73{,}500.00 = 209.95.$$

Clearly, 209.95 is very close to the 210 that we had obtained earlier.

Example 5.16

A company making toy sloths found the cost function (in dollars) for making x toy sloths a week to be $C(x) = 3000 + 150x - 0.075x^2$. Use the marginal cost function to estimate the cost of making the 501st toy sloth. Then find the exact cost of making the 501st toy sloth.

Solution: First we find the marginal cost function by taking the derivative of the cost function,

$$C'(x) = 150 - 0.15x.$$

We estimate the cost of making the 501st toy sloth by finding the marginal cost of 500,

$$C'(500) = 150 - 0.15(500) = 75.$$

So the cost of making the 501st toy sloth is approximately 75 dollars. To find the exact cost of making the 501st toy sloth we first find the exact cost of making 501 and 500 toy sloths;

$$C(501) = 3000 + 150(501) - 0.075(501)^2 = 59{,}324.925$$
$$C(500) = 3000 + 150(500) - 0.075(500)^2 = 59{,}250.$$

Thus, the exact cost of making the 501st toy sloth is given by

$$C(501) - C(500) = 59{,}324.925 - 59{,}250 = 74.925 \approx 74.93.$$

Of course, business people can also talk about **marginal revenue** and **marginal profit** as well. Marginal revenue is defined to be the additional revenue obtained by producing (and presumably selling) one additional unit of a product or service. Marginal profit is the additional profit obtained by producing (and presumably selling) one additional unit of a product or service. The **marginal revenue function** is simply the derivative of the revenue function and the **marginal profit function** is simply the derivative of the profit function.

Marginal Revenue, Cost, and Profit Functions:

If x is the quantity produced then:

$R'(x)$	Marginal Revenue Function
$C'(x)$	Marginal Cost Function
$P'(x) = R'(x) - C'(x)$	Marginal Profit Function

Example 5.17

The total revenue in dollars from the sales of widgets is given by $R(x) = 15000x - 120x^2$. Use the marginal revenue to approximate the revenue from the sale of the 36th widget. Then find the exact revenue obtained from selling the 36th widget.

Solution: First we find the marginal revenue function,

$$R'(x) = 15000 - 240x.$$

To estimate the marginal revenue from the sale of the 36th widget we substitute in 35,

$$R'(35) = 15000 - 240(35) = 6600.$$

So we expect the sale of the 36th widget to make an additional 6600 dollars. To find the exact additional revenue that the sale of the 36th widget makes we first find

$$R(36) = 15000(36) - 120(36)^2 = 384{,}480$$
$$R(35) = 15000(35) - 120(35)^2 = 378{,}000.$$

Thus, the exact revenue from selling the 36th widget is given by

$$R(36) - R(35) = 384{,}480 - 378{,}000 = 6480.$$

The Price-Demand Function

The **price-demand function** is a very important concept in business and economics that relates the demand of an item to its price. Price-demand functions are also often simply called demand functions. As a general rule of thumb, the more items that are produced, the lower the price per item will be. For example, luxury goods are usually expensive in part because they are produced in limited quantities, which makes the items seem desirable. But if these items were mass-produced, their prices would naturally fall due to the increased availability.

The price-demand function is a function that tries to capture that relationship. Some price demand functions are estimated by marketing departments or found using economic theory. In this book, we will always assume a price-demand function is known and is given to us. We will not concern ourselves with how they are derived. Now, suppose we were given the price-demand function

$$p = 50 - 0.01x.$$

The variable p stands for the price (in dollars) and the variable x stands for the number of items available to be sold. So, suppose that we have $x = 10$ items to sell, then the price we can charge per item is given by

$$p = 50 - 0.01(10) = 49.90.$$

So, when we only have 10 items to sell the price is 49.9 dollars per item. Now suppose we had $x = 100$ items to sell. Then the price we can charge per item is given by

$$p = 50 - 0.01(100) = 49.$$

When we have 100 items to sell the price goes down a little to 49 dollars per item. Now suppose we had $x = 1000$ items to sell. In this case the price per item we can charge is given by

$$p = 50 - 0.01(1000) = 40.$$

The more items we have to sell, the less we are able to charge for the item, because a greater availability means people are not willing to pay as much for each item. This is simply how supply and demand usually work. Thus, price-demand functions are decreasing functions. Finally, very often price-demand functions are given in terms of p for price and q for quantity instead of in terms of x. The above price demand function would simply be written as $p = 50 - 0.01q$. Do not let this confuse you; you can easily change the variable q to x if you want. The name of the variable does not matter. Also, we generally assume the price is in terms of dollars.

Example 5.18

Suppose a price demand function is given by $p = 75 - 0.02x$. What price can you charge if you have $x = 1000$ items to sell.

Solution: We simply have to substitute 1000 into the x to get

$$p = 75 - 0.02(1000) = 55.$$

Thus, price we can charge when $x = 1000$ is 55 dollars.

The useful thing about price-demand functions is that we can use them to write down the revenue function. First of all, what is revenue? Revenue is the total amount of money you get from selling some items. Suppose you sold 50 items for 10 dollars each. Then the revenue would be $50 \times 10 = 500$ dollars. If you sold 500 items at 8 dollars each then your revenue would be $500 \times 8 = 4000$ dollars. The general formula is given as

$$\text{revenue} = \text{price} \times \text{quantity}.$$

This is often written as

$$R(q) = p \times q$$

or as

$$R(x) = p \times x.$$

Suppose we had the same price-demand function as before, $p = 50 - 0.01x$. If we were asked to find the revenue functions, then we would simply multiply the price-demand function by the quantity x,

$$R(x) = p \times x$$

$$= (50 - 0.01x)x$$
$$= 50x - 0.01x^2.$$

Notice, the revenue function is given in terms of the variable x, the quantity of items sold. The variable p no longer appears.

Example 5.19

A company making widgets finds that the price-demand function for the widgets is given by $p = 1500 - 0.01x$. What is the revenue function for the widgets?

Solution: To find the revenue function we simply have to multiply the price by the quantity x.

$$R(x) = \text{price} \times \text{quantity}$$
$$= (1500 - 0.01x)x$$
$$= 1500x - 0.01x^2.$$

Example 5.20

Suppose a company that makes some product finds the price demand function given by $x = 1500 - 20p$. What is the revenue function?

Solution: Be careful, here the price-demand function gives x in terms of p and not the other way around. To solve this we first have to write p in terms of x. In other words, we need to solve for p.

$$x = 1500 - 20p \implies 20p = 1500 - x$$
$$\implies p = \frac{1500 - x}{20}$$
$$\implies p = 75 - \frac{x}{20} = 75 - 0.05x.$$

Once we have solved for p we can find the revenue function just like before. When our functions are not complicated, our revenue function often ends up being a quadratic equation.

$$R(x) = p \times x$$
$$= (75 - 0.05x)x$$
$$= 75x - 0.05x^2.$$

Example 5.21

Suppose the price-demand function for some product is given by $p = 300 - 0.01x$. What is the instantaneous rate of change in revenue when production is 10,000?

Solution: First we find the revenue function using the price-demand function,

$$R(x) = (300 - 0.01x)x$$
$$= 300x - 0.01x^2.$$

Then we need to find the derivative of the revenue function,

$$R'(x) = 300 - 0.02x.$$

This is also called the marginal revenue function. We then substitute 10,000 in for x to get

$$R'(10,000) = 300 - 0.02(10,000) = 100.$$

Thus, the instantaneous rate of change in revenue when production is 10,000 is 100. This means that when sales increase from 10,000 units to 10,001 units the revenue increases by 100 dollars.

Example 5.22

Suppose you know the price-demand function for some product is given (in thousands of dollars) by $p = 200 - 3q$. Find the marginal revenue when the production is 10 units. What does this number represent?

Solution: Writing quantity using the variable x instead of q, we have

$$R(x) = (200 - 3x)x$$
$$= 200x - 3x^2.$$

We then take the derivative of the revenue function to get the marginal revenue function,

$$R'(x) = 200 - 6x,$$

which we then use to find the marginal revenue when production is 10,

$$R'(10) = 200 - 6(10) = 140.$$

The additional revenue, if production increased from 10 to 11 units, is 140,000 dollars.

Example 5.23

Suppose the price-demand function for an item was given by $p = 1000 - 0.1x$. What is the total revenue when 3000 items are sold? What is the marginal revenue when sales are at 3000?

Solution: To answer this question we need to first know the revenue function,
$$R(x) = (1000 - 0.1x)x$$
$$= 1000x - 0.1x^2.$$

We can now substitute in $x = 3000$ to find the revenue when sales are 3000,
$$R(3000) = 1000(3000) - 0.1(3000)^2 = 2,100,000.$$

So, when we sell 3000 items our revenue is 2,100,000 dollars. To find the marginal revenue we now need to take the derivative of the revenue function to get $R'(x) = 1000 - 0.2x$. When $x = 3000$, we have

$$R'(3000) = 1000 - 0.2(3000) = 400.$$

The marginal revenue when sales are at 3000 is 400. Thus, when sales increase from 3000 to 3001 the approximate increase in revenues is 400.

Revenue and Profit Maximization

In economics we often talk about **monopoly regime** where a monopolist controls the demand and the price in order to maximize profit. Profit is maximized when
$$P'(x) = 0 \quad \text{and} \quad P''(x) < 0.$$
As we have seen before $P(x) = R(x) - C(x)$. Using this identity we find that the first and second derivatives of the profit function are
$$P'(x) = R'(x) - C'(x) \quad \text{and} \quad P''(x) = R''(x) - C''(x).$$
As we have seen in the previous chapter, the derivative of the total revenue is the marginal revenue, and the derivative of the total cost is the marginal cost, i.e., $R'(x) = MR$ and $C'(x) = MC$. Solving $P'(x) = 0$ yield the following identity:
$$P'(x) = 0 \implies R'(x) - C'(x) = 0$$
$$\implies R'(x) = C'(x)$$
$$\implies MR = MC.$$

So, the first condition $P'(x) = 0$ is equivalent to saying that $MR = MC$, which means that profit is maximized when the marginal revenue is equal to marginal cost.

Example 5.24

The price-demand function of a commodity is given by the function $p = 11 - 6x$ and its total cost is given by $C(x) = x^3 - 3x^2 + 2x - 50$.

1. Find the marginal cost function.
2. Find the profit function.
3. Find the maximum profit.

Solution:

1. The marginal cost is defined as

$$MC = C'(x) \implies MC = 3x^2 - 6x + 2.$$

2. To find the profit function we need to first find the revenue function $R(x)$,

$$R(x) = (11 - 6x)x$$
$$= 11x - 6x^2.$$

   The profit function is defined as $P(x) = R(x) - C(x)$, therefore

$$P(x) = 11x - 6x^2 - (x^3 - 3x^2 + 2x - 50)$$
$$= -x^3 - 3x^2 + 9x + 50.$$

3. To find the maximum profit, we compute $P'(x) = -3x^2 - 6x + 9$ and then solve $P'(x) = 0$ to find the stationary points:

$$P'(x) = 0 \implies -3x^2 - 6x + 9 = 0$$
$$\implies -3(x^2 + 2x - 3) = 0$$
$$\implies -3(x + 3)(x - 1) = 0$$
$$\implies x = -3 \text{ or } x = 1.$$

Since a negative quantity has no meaning in economics, we only consider the second solution. We determine the nature of the point $x = 1$ by choosing two test values as follows:

$x = 0$	$x = 2$
$P'(0) = 9$	$P'(0) = -15$
positive slope	negative slope
$\Rightarrow$ increasing	$\Rightarrow$ decreasing

Therefore, there is a local maximum at $x = 1$, and the maximum profit is $P(1) = 55$.

Example 5.25

Given a price-demand function $p = 12 - 3x$, a monopolist has a total cost function given by $C(x) = 2x^3 - 6x^2 + 12x - 20$. Find the maximum profit for this monopolist.

Solution: We first find the total revenue function, $R(x)$,

$$R(x) = (12 - 3x)x$$
$$= 12x - 3x^2.$$

The profit function, $P(x)$, will then be

$$P(x) = R(x) - C(x)$$
$$= 12x - 3x^2 - (2x^3 - 6x^2 + 12x - 20)$$
$$= 12x - 3x^2 - 2x^3 + 6x^2 - 12x + 20$$
$$= -2x^3 + 3x^2 + 20$$
$$\implies P'(x) = -6x^2 + 6x$$
$$\implies P''(x) = -12x + 6.$$

To find the maximum profit we first need to find the quantity that maximizes the profit, which occurs when the derivative of the profit function is equal to zero.

$$P'(x) = 0$$
$$\implies -6x^2 + 6x = 0$$
$$\implies -6x(x - 1) = 0$$
$$\implies x = 0 \text{ or } x - 1 = 0$$
$$\implies x = 0 \text{ or } x = 1.$$

We then use the second derivative test to determine if these points correspond to minimums or maximums.

$x = 0$	$x = 1$
$P''(0) = 6 > 0$	$P'(0) = -6 < 0$
concave up	concave down
$\Rightarrow x$ is a minimum	$\Rightarrow x$ is a maximum

To find the maximum profit we substitute $x = 1$ in the profit function: $P(1) = 21$.

Example 5.26

A company sells x used business books every year at p dollars each. The price-demand function for this company is given by: $p = 18 - 0.002x$.

1. How many used business books does the company have to sell to reach the maximum revenue?
2. What will be the maximum revenue?
3. What price would the company have to charge to make maximum revenue?

Solution:

1. We first find the total revenue function, $R(x)$,

$$R(x) = (18-0.002x)x$$
$$= 18x - 0.002x^2.$$

To find the maximum revenue we must find the stationary points by solving
$$R'(x) = 0$$
$$\Longrightarrow 18 - 0.004x = 0$$
$$\Longrightarrow x = 4500.$$

If the second condition holds, to determine whether this is a maximum or a minimum, we will check the second condition holds, that is, that $R''(x) < 0$ for $x = 4500$. But $R''(x) = -0.004$, which is negative for all values of x, therefore there is a maximum at $x = 4500$. The maximum revenue will be reached when the quantity sold is 4500 business books per year.

2. To find the maximum revenue we substitute $x = 4500$ in the revenue function:

$$R(4500) = 18(4500) - 0.002(4500)^2$$
$$= 40{,}500.$$

Therefore, the maximum revenue the company can achieve is 40,500 dollars per year.

3. To find the price of the business books when revenue is a maximum, we will substitute $x = 4500$ in the demand function:

$$p = 18 - 0.002(4500) = 9.$$

Therefore, to maximize revenue the company should sell the used business books at 9 dollars each.

Example 5.27

Suppose a manufacturer of widgets sells x widgets each day for p dollars each. The price-demand function is given by $p = 60 - 0.05x$. How many widgets must the manufacturer sell to maximize their revenue? What would the maximum revenue be? And what price would the manufacturer have to charge to make maximum revenue?

Solution: First, we have to use the price-demand function to find the revenue function,

$$R(x) = (60 - 0.05x)x$$
$$= 60x - 0.05x^2.$$

Now, in order to find the number of widgets that maximize revenue we take the derivative of the revenue function and set it equal to zero and solve for x,

$$R'(x) = 0$$
$$\implies \quad 60 - 0.1x = 0$$
$$\implies \quad x = 600.$$

Technically we should check to see if $x = 600$ is a maximum, but we will skip that step right now. The maximum revenue is obtained by evaluating the revenue function at $x = 600$,

$$R(600) = 60(600) - 0.05(600)^2 = 35{,}820.$$

Thus the maximum revenue is 35,820 dollars each day. To find out the price the manufacturer would have to sell the widgets for we evaluate the price-demand function at $x = 600$,

$$p = 60 - 0.05(600) = 30.$$

So in order for the manufacturer to maximize revenue they should sell the widgets for 30 dollars.

DIMINISHING MARGINAL RETURN

Economists and business people often talk about points of diminishing marginal returns, called diminishing returns for short. The point of **diminishing returns** occurs when the marginal returns stop increasing and start decreasing. In other words, when the marginal returns start diminishing. This is exactly the same thing as saying that the slopes of the tangent lines stop increasing and start decreasing, which is equivalent to saying that the graph goes from concave up to concave down. Thus, the point of diminishing returns happens at a point of inflection.

Suppose a company estimates that it will sell $N(x)$ widgets a week if it spends x thousand dollars on advertising each week, where

$$N(x) = -0.25x^3 + 6x^2 + 8x,$$

for $0 \leq x \leq 16$. This is graphed below.

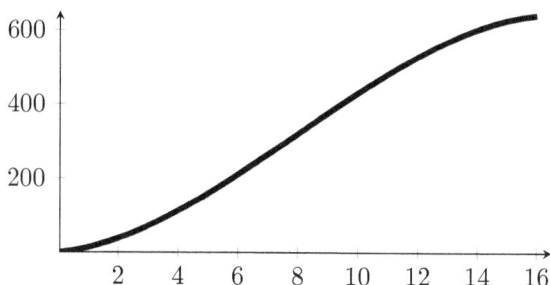

How much should the company spend on advertising? If the advertising is too little there will not be many sales. But spending more than 16 thousand a week on advertising does not increase sales much at all. In fact, if an increase in advertising only results in a very small increase in sales then there is no reason to increase the money spent on advertising. So how do companies decide how much to spend on advertising?

As advertising increases from 2 to 3 thousand to then the sales increases by

$$N(3) - N(2) = 71.25 - 38 = 33.25 \approx 33.$$

Whereas, as advertising increases from 15 to 16 thousand to then the sales increases by

$$N(16) - N(15) = 640 - 626.25 = 13.75 \approx 14.$$

So, a one thousand dollar increase in advertising from two thousand to three thousand results in an increase in sales of 33 widgets while a one thousand dollar increase in advertising from 15 thousand to 16 thousand only results in an increase in sales of 14 widgets. At some point the changes in sales stop increasing and start decreasing. This point is called the **point of diminishing returns**. This is the point many companies use to choose to stop spending money on advertising. Note that nothing says a company has to stop spending on advertising at this point, but this point is often used in practice. The point of diminishing returns is an inflection point. We can find inflection points by taking the second derivative, setting it equal to zero, and solving for x.

$$N''(x) = -1.5x + 12 = 0 \implies -1.5x + 12 = 0$$
$$\implies x = 8.$$

Thus the point of diminishing returns is at $x = 8$. We plot this point on the graph. To the left of this point the graph is concave up and to the right of this point the graph is concave down.

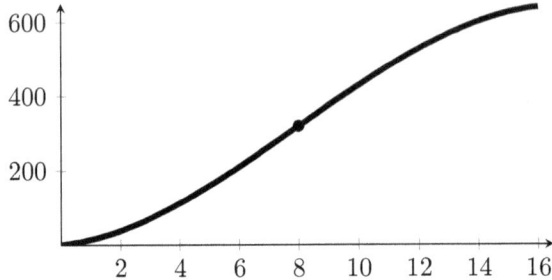

Example 5.28

A company producing widgets wants to expand its workforce. The company estimates it will produce $A(x)$ widgets a day by hiring x workers when $0 < x < 30$. How many workers will there be at the point of diminishing returns if $A(x) = \frac{-1}{6}x^4 + \frac{20}{3}x^3$. How many widgets will be produced a day at this point?

Solution: Finding the point of diminishing returns means finding the point of inflection. First we need to find the second derivative,

$$A''(x) = -2x^2 + 40x.$$

Next, we set the second derivative equal to zero and solve for x,

$$\begin{aligned} A''(x) = 0 &\implies -2x^2 + 40x = 0 \\ &\implies -2x(x - 20) = 0 \\ &\implies x = 0 \quad \text{and} \quad x = 20. \end{aligned}$$

Clearly the $x = 20$ solution is the one we are interested in. Technically we should check that there is a change in concavity from concave up to concave down, but we will forgo that here. There are 20 workers at the point of diminishing returns. To find the number of widgets that will be produced at this point we evaluate the function at $x = 20$,

$$A(20) = \frac{-1}{6}(20)^4 + \frac{20}{3}(20)^3 = 26{,}666.67 \approx 26{,}667.$$

Since we are talking about the number of widgets, it makes sense to round to the nearest whole number.

Example 5.29

A new bakery shop starts by employing three staff who can service about three customers per minute. After a few months, the owner decides to employ more staff to improve the service and make more profit. The relationship between the number of staff employed to the number of customers serviced per minute is given by:

$$N(x) = -0.0075x^4 + 0.12x^3 \qquad 0 \leq x \leq 12$$

At what point will the law of diminishing returns come into play? What should the bakery owner do at this point?

Solution: Finding the point of diminishing returns means finding the point of inflection. First, we need to find the second derivative,

$$N'(x) = -0.03x^3 + 0.36x^2$$
$$N''(x) = -0.09x^2 + 0.72x.$$

Next, we set the second derivative equal to zero and solve for x,

$$N''(x) = 0 \implies -0.09x^2 + 0.72x = 0$$
$$\implies -x(0.09x - 0.72) = 0$$
$$\implies x = 0 \quad \text{and} \quad x = 8.$$

It is clear that $x = 8$ is the solution we are interested in. Again, technically we should check that there is a change in concavity from concave up to concave down, but again, we will forgo that here. There are 8 workers at the point of diminishing returns. At this point, the owner should stop hiring new staff.

5.5 PROBLEMS

Question 5.1 *Find the stationary points for the following functions and determine their nature.*

(a) $f(x) = -x^2 + 5$

(b) $f(x) = 4x^2 - 16x + 7$

(c) $f(x) = -x^2 - 14x + 32$

(d) $f(x) = x^2 + 7x - 6$

(e) $f(x) = x^3 - 12x - 4$

(f) $f(x) = x^3 - x^2 + 5$

(g) $f(x) = -x^3 + 3x^2 - 2$

(h) $f(x) = x^3 - 3x^2 + 1$

(i) $f(x) = x^4 - 8x^2 + 6$

(j) $f(x) = -x^4 - x^3 + 3x^2 + 5x + 2$

(k) $f(x) = -x^4 - 5x^3 - 6x^2 + 4x + 8$

(l) $f(x) = x^4 + 6x^3 - 54x - 81$

Question 5.2 *Find the points of inflection for the following functions:*

(a) $f(x) = -x^3 + 9x + 3$

(b) $f(x) = -\frac{2}{3}x^3 + 6x^2 - x$

(c) $f(x) = 10x^3 - 3x^5$

(d) $f(x) = x^4 - 24x^2$

(e) $f(x) = 2x^3 - 9x^2 + 12x$

(f) $f(x) = 7x - x^3$

(g) $f(x) = 7x^3 + 2x + 7$

(h) $f(x) = \frac{4}{3}x^3 - 12x^2 + 10x + 45$

(i) $f(x) = x^3 + 3x^2 - x - 24$

(j) $f(x) = x^3 + 7x + 1$

(k) $f(x) = x^3 - 12x^2 + 2x + 15$

(l) $f(x) = \frac{x^4}{2} - 2x^3 + 12$

Question 5.3 *Find the absolute extrema of the following functions on each of the given domains D_1 and D_2.*

(a) $f(x) = x^3 - 12x - 4 \quad D_1 = [0, 4], \quad D_2 = [-3, 3]$

(b) $f(x) = x^3 - x^2 + 5 \quad D_1 = [-1, 2], \quad D_2 = [0, 2]$

(c) $f(x) = 2x^3 - 3x^2 \quad D_1 = [-1, 1], \quad D_2 = [0, 2]$

(d) $f(x) = x^3 + 3x \quad D_1 = [0, 2], \quad D_2 = [-2, 1]$

Question 5.4 *Find the stationary points of the function $f(x) = 2x^3 - 15x^2 + 24x + 6$ and determine whether they are maxima or minima.*

Question 5.5 *The cost function for some item is given by $C(x) = -0.095x^2 + 150x + 3000$. Find the marginal cost of producing 600 units. Find the exact cost of producing the 601st unit.*

Question 5.6 *The cost function for a company producing widgets was esti-mated to be $C(x) = 200{,}000 + 500x$. Find the marginal cost of making 500 widgets and estimate the exact cost for making the 501st widget.*

Question 5.7 *The cost function for producing some item is given by $C(x) = -0.025x^2 + 30x + 1500$. Find the exact cost of producing the 31st item. Find the marginal cost when production is 30. Compare the results.*

Question 5.8 *A company making widgets has the following price-demand function $x = 6000 - 800p$ where p is the price per widget. Find the marginal revenue at production levels of 1000 widgets, 3000 widgets, and 5000 widgets.*

Question 5.9 *The price-demand function for a certain good is $p = 350 - 5x$ where x is the number of items and p is the price per item. Estimate the change in revenue when production increases from 10 items to 11 items using the marginal revenue function. Then calculate the exact change in revenue.*

Question 5.10 *Suppose a manufacturer sells x widgets every day for p dol-lars each. The price-demand function is given by $p = 60 - 0.05x$. The daily cost function is given by $C(x) = 4000 + 15x$. How many widgets need to be sold to maximize the profit? What is the maximum profit? What price should the manufacturer charge per widget to maximize the profit?*

Question 5.11 *A company produces and sells x widgets a week. The weekly price-demand function is given by $p = 200 - 0.1x$. The weekly cost function is $C(x) = 15000 + 95x$. How many widgets does the company need to make to maximize their profits? What is the maximum weekly profit? What price should the company charge in order to maximize their profits?*

Question 5.12 *A company produces and sells x items a week. The weekly price-demand function is given by $p = 300 - 0.4x$. The weekly cost function is $C(x) = 4000 + 150x$. How many items does the company need to make to maximize their profits? What is the maximum weekly profit? What price should the company charge in order to maximize their profits?*

Question 5.13 *A company buys and stores items in bulk. If the item has the cost function $C(x) = \frac{4,000,000}{x} + 0.25x$, how many items should the company buy at a time to minimize the cost?*

Question 5.14 *A company buys and stores widgets in bulk. If the widgets have the cost function $C(x) = \frac{144}{x} + 0.0001x$, how many widgets should the company buy at a time to minimize the cost?*

Question 5.15 *The revenue for a company that makes widgets is given by the revenue function $R(x) = 500x - 10x^2$ where x is the number of widgets produced and the revenue is in thousands of dollars. What is the average rate of change in revenue with respect to widgets produced as production increases*

 (a) *from 25 to 30,* (b) *from 30 to 40,* (c) *from 35 to 40.*

Question 5.16 *Suppose the cost of producing x widgets in dollars is given by the cost function $C(x) = 500 + 15x - 0.2x^2$. Find the marginal cost of producing 30 widgets.*

Question 5.17 *Suppose that the total profit in hundreds of dollars from selling x items is given by the profit function $P(x) = -x^2 + 10x - 30$. Find the marginal profit from selling 4 items.*

Question 5.18 *Suppose the revenue function for a certain product is given by $R(x) = 350x - 0.75x^2$.*

 (a) *What is the actual increase in revenue as production increases from 150 units to 200 units?*

 (b) *What is the average rate of change in revenue as production increases from 150 units to 200 units?*

 (c) *What is the marginal revenue when production is at 150 units? What does this mean?*

Question 5.19 *A company making a certain product finds that the price-demand function for the product is given by* $p = 35 - 0.05x$.

(a) *Find the revenue function for the product.*

(b) *Find the actual change in revenue if production increases from* 200 *units to* 300 *units.*

(c) *Find the average rate of change in revenue over this same interval.*

(d) *Find the revenue when production is* 200 *units.*

(e) *Find the instantaneous rate of change of revenue when production is* 200 *units.*

Question 5.20 *A company making a certain product has the following total cost function* $C(x) = 3x^3 - 6x^2 - 8x$, *where* x *indicates the quantity.*

(a) *Find the marginal cost function for the product.*

(b) *Find the average cost of the product.*

(c) *Find the local extrema of the total cost function.*

Question 5.21 *The demand for a certain commodity is* $p = 8 - x$, *where* p *is the price and* x *is the quantity.*

(a) *Find the price and quantity for which the revenue is maximum.*

(b) *Represent graphically the demand, total revenue, and marginal revenue on the same graph.*

Question 5.22 *A copy machine producer produces q copy machine per month with a total cost of* $C(q) = \frac{q^2}{36} + 3q + 120$ *dollars. If the producer is a monopolist with a demand function of* $q = 60 - 2p$, *where* p *is the price of a copy machine.*

(a) *Find the number of copy machines the producer should produce per month to maximize its profit.*

(b) *What is the maximum profit?*

(c) *What is the monopolist price?*

Question 5.23 *A small shop manufacturing specialty gift boxes is planning to expand its workforce. It estimates that the number of gift boxes produced by hiring* x *new workers is given by:*

$$H(x) = -0.25x^4 + 5x^3 \qquad\qquad 0 \le x \le 15$$

At what point does the point of diminishing returns come into play? What should the shop owner do at this point?

Question 5.24 *A publishing company estimates that it will sell* $N(x)$ *units of a product if it spends* x *(thousand dollars) on advertising using the function:*

$$N(x) = -0.06x^3 + 18x^2 + 12x + 600 \qquad\qquad 0 \le x \le 200$$

1. *Calculate the point of diminishing returns. What is the amount being spent on advertising at this point?*

2. *What should the company do at this point?*

CHAPTER 6

Antiderivatives and Integration

Here we begin by giving antiderivative rules for a number of simple functions before then exploring the geometric meaning of the antiderivative and giving a formal definition. We then present the fundamental theorem of calculus, which allows us to compute definite integrals before finally looking at a variety of business and economics applications.

6.1 ANTIDERIVATIVES

So far in this book, we have discussed differentiation, or finding derivatives. However, sometimes we need to go backwards. Given the derivative of a function we want to find the original function. The original function would be called the **antiderivative** function and the process of finding the antiderivative function is called **integration**.

Consider the function $f_1(x) = 40x$. It is clear that $f_1'(x) = 40$. In fact, consider the following functions:

$$f_2(x) = 40x + 10 \qquad f_3(x) = 40x - 20 \qquad f_4(x) = 40x + \pi$$
$$f_2'(x) = 40 \qquad\qquad f_3'(x) = 40 \qquad\qquad f_4'(x) = 40$$

and so on. In fact, if c is any constant number at all, then we have

$$f(x) = 40x + c,$$
$$f'(x) = 40.$$

So, many different functions have the derivative $f'(x) = 40$. In fact, any function of the form $f(x) = 40x + c$, where c is a constant number, has the derivative $f'(x) = 40$. So if we wanted to find the antiderivative of $f'(x) = 40$ which of these functions would it be? We do not know unless we are given extra information. If we are not given any extra information then all we can

DOI: 10.1201/9781003480235-6

say is that the antiderivative of $f'(x) = 40$ is $f(x) = 40x + c$, where c is an unknown constant number.

Antiderivative of $f(x)$

The antiderivative of a function $f(x)$ is a function $F(x)$ whose derivative is $f(x)$. Symbolically, we write

$$F'(x) = f(x)$$

or

$$\frac{d}{dx}F(x) = f(x)$$

Example 6.1

Find the antiderivative of the function $f(x) = 3$.

Solution: Essentially, we need to think of a function whose derivative is 3. It should be obvious that $F(x) = 3x$ would work. Indeed, using our differentiation rules, we have

$$\frac{d}{dx}F(x) = \frac{d}{dx}(3x)$$
$$= 3 \cdot \frac{d}{dx}(x)$$
$$= 3 \cdot 1$$
$$= 3.$$

However, it is important to note that this is just one possible antiderivative of $f(x) = 3$. In general, if $F(x)$ is an antiderivative of $f(x)$, then any function of the form $F(x) + c$ (where c is an arbitrary constant) is also an antiderivative of $f(x)$. Therefore, we can also write the antiderivative of $f(x) = 3$ as $F(x) = 3x + c$, where c is an arbitrary constant.

In essence, finding the antiderivative of a function $f(x)$ is like finding the derivative, only "going backwards" and then adding a constant c to the result. So, at least for our simple derivative rules, we can find the corresponding antiderivatives rules simply by "going backwards." But before we do that we need to introduce a little additional terminology and notation.

Indefinite Integral of $f(x)$

Suppose that $F(x)$ is an antiderivative of the function $f(x)$. We would say that the **indefinite integral** of $f(x)$ is $F(x) + c$ and would write

$$\int f(x)\, dx = F(x) + c.$$

The symbol $\int$ is called an **integral sign**, the function $f(x)$ is called the **integrand**, and the dx indicates that the variable of integration is x.

We will discuss the meaning and reasons behind this notation in a few sections, but for now we will simply recognize it as a different notational convention that is used to represent antiderivatives. In fact, most antiderivative rules are written using this notation, which is why we introduced it here. The rest of this section will simply be a listing of the various antiderivative, or indefinite integral, rules associated with many of the derivative rules we have learned. The first antiderivative rule, illustrated in the last example, is the integral of constant functions.

Indefinite Integral of a constant function:

For a constant k,

$$\int k\, dx = kx + c,$$

where c is an arbitrary constant.

For the remainder of the rules we shall omit the phrase "where c is an arbitrary constant."

Example 6.2

Find the antiderivative of the function $f(x) = 7$.

Solution: Using the above rule, it is easy to see that

$$\int 7\, dx = 7x + c.$$

It is trivial to double check this by simply taking the derivative of $7x + c$, which gives us 7.

Example 6.3

Find the antiderivative of the function $f(x) = -4$.

Solution:
$$\int -4\, dx = -4x + c.$$

Indefinite Integral Power Rule:

For any real number $n \neq -1$,

$$\int x^n\, dx = \frac{x^{n+1}}{n+1} + c.$$

We will deal with the case when $n = -1$ below.

Example 6.4

Find the antiderivative of $f(x) = x^3$.

Solution:
$$\int x^3\, dx = \frac{x^{3+1}}{3+1} + c = \frac{x^4}{4} + c$$

Example 6.5

Find the indefinite integral of $f(x) = x^{-5}$.

Solution:
$$\int x^{-5}\, dx = \frac{x^{-5+1}}{-5+1} + c = -\frac{x^{-4}}{4} + c$$

Example 6.6

Find the indefinite integral of $f(x) = \frac{1}{\sqrt{x}}$.

Solution: It is useful to write the integrand in "power rule friendly form."

$$\int \frac{1}{\sqrt{x}}\, dx = \int x^{-\frac{1}{2}}\, dx = \frac{x^{-\frac{1}{2}+1}}{-\frac{1}{2}+1} + c = \frac{x^{\frac{1}{2}}}{\frac{1}{2}} + c = 2x^{\frac{1}{2}} + c = 2\sqrt{x} + c$$

Note the two different ways this question could be phrased: you could be asked to find either the antiderivative or the indefinite integral of a function $f(x)$. The next two indefinite integral rules allow us to find antiderivatives in a wide variety of situations.

Indefinite Integral Constant Multiple Rule and Sum/Difference Rule:

Suppose k is a constant and $f(x)$ and $g(x)$ are functions whose integrals exist. Then

$$\int k f(x)\, dx = k \int f(x)\, dx$$

and

$$\int \left(f(x) \pm g(x) \right) dx = \int f(x)\, dx \pm \int g(x)\, dx.$$

Example 6.7

Find the indefinite integral of $f(x) = 6x^2$.

Solution: Using the constant multiple rule, we have

$$\int 6x^2\, dx = 6 \int x^2\, dx \qquad \text{Constant Multiple Rule}$$

$$= 6 \left[\frac{x^{2+1}}{2+1} + c \right] \qquad \text{Indefinite Integral Power Rule}$$

$$= 6 \cdot \frac{x^3}{3} + 6 \cdot c \qquad \text{Distribution}$$

$$= 2x^3 + c. \qquad \text{Rewrite}$$

In order to illustrate a point, in this example, we did something we will not do again. When we distributed the 6 we wrote down $6 \cdot c$ for the second term. The letter c represents a constant, which is just any arbitrary unknown number. But six times an arbitrary unknown number is simply another arbitrary unknown number, which we simply call c again.

Example 6.8

Find the indefinite integral of $f(x) = 4x^{-5}$.

Solution:
$$\int 4x^{-5}\, dx = 4\frac{x^{-5+1}}{-5+1} + c$$
$$= 4\frac{x^{-4}}{-4} + c$$
$$= -x^{-4} + c$$
$$= \frac{-1}{x^4} + c$$

Example 6.9

Find the indefinite integral of $f(x) = 4x^2 + 2x$.

Solution: Here we will use the sum/difference rule. We are given the sum of two functions. We can take the integral of this sum by taking the integral of each term and then summing them.

$$\int f(x)\, dx = \int \left(4x^2 + 2x\right) dx$$
$$= \int 4x^2\, dx + \int 2x\, dx$$
$$= 4\int x^2\, dx + 2\int x\, dx$$
$$= 4 \cdot \frac{x^3}{3} + 2 \cdot \frac{x^2}{2} + c$$
$$= \frac{4x^3}{3} + x^2 + c$$

Example 6.10

Find the indefinite integral of $f(x) = \frac{8}{5}x^3 + 3x^2 - 5x - 10$.

Solution: Finding the indefinite integral of this function will require all the rules listed so far, including multiple uses of the sum/difference rule.

$$\int f(x)\, dx = \int \left(\frac{8}{5}x^3 + 3x^2 - 5x - 10\right) dx$$

$$= \int \frac{8}{5} x^3 \, dx + \int 3x^2 \, dx - \int 5x \, dx - \int 10 \, dx$$

$$= \frac{8}{5} \frac{x^4}{4} + 3 \frac{x^3}{3} - 5 \frac{x^2}{2} - 10x + c$$

$$= \frac{2x^4}{5} + x^3 - \frac{5x^2}{2} - 10x + c$$

Of course, as you gain familiarity with the rules you will not need to write down every line of the above calculation explicitly.

Indefinite Integral Exponential Rules:

For any base $b > 0, b \neq 1$,

$$\int b^x \, dx = \frac{b^x}{\ln(b)} + c.$$

Given the base e, we have

$$\int e^x \, dx = e^x + c.$$

Example 6.11

Find the antiderivative of $f(x) = 5^x$.

Solution:

$$\int 5^x \, dx = \frac{5^x}{\ln(5)} + c$$

Example 6.12

Find the antiderivative of $f(x) = 3e^x$.

Solution:

$$\int 3e^x \, dx = 3 \int e^x \, dx = 3e^x + c$$

Example 6.13

Find the indefinite integral of $A(t) = 1000(1.025)^t$.

Solution:

$$\int 1000(1.025)^t \, dt = 1000 \int (1.025)^t \, dt$$

$$= 1000 \frac{(1.025)^t}{\ln(1.025)} + c$$

$$= \frac{1000(1.025)^t}{\ln(1.025)} + c$$

Example 6.14

Find the antiderivative of $f(x) = 3^x + x^3 - 3$.

Solution:

$$\int 3^x + x^3 - 3 \, dx = \int 3^x \, dx + \int x^3 \, dx - \int 3 \, dx$$

$$= \frac{3^x}{\ln(3)} + \frac{x^4}{4} - 3x + c$$

While, in one sense, this is a very easy integral, it is essential to understand how each of the three terms differ. Students often try to apply the wrong rule in different situations.

Indefinite Integral Rule for $f(x) = \frac{1}{x}$:

Given the function $f(x) = \frac{1}{x}$,

$$\int \frac{1}{x} \, dx = \ln|x| + c.$$

Recall, we could also write $\frac{1}{x}$ as x^{-1}.

Example 6.15

Find the antiderivative for $f(x) = \frac{2}{3x}$.

Solution:

$$\int \frac{2}{3x} \, dx = \frac{2}{3} \int \frac{1}{x} \, dx = \frac{2}{3} \ln|x| + c$$

Indefinite Integral Logarithm Rules:

For any base $b > 0$, $b \neq 1$,

$$\int \log_b(x)\, dx = x \log_b(x) - \frac{x}{\ln(b)} + c.$$

Given the base e, we have

$$\int \ln(x)\, dx = x \ln(x) - x + c.$$

Example 6.16

Show the rule $\int \ln(x)dx = x\ln(x) - x + c$ is true by taking the derivative of $x\ln(x) - x + c$.

Solution: Note that taking the derivative of the first term requires the use of the product rule.

$$\left[x\ln(x) - x + c\right]' = \left[x\right]' \cdot \left[\ln(x)\right] + \left[x\right] \cdot \left[\ln(x)\right]' - 1 + 0$$

$$= \ln(x) + x \cdot \frac{1}{x} - 1$$

$$= \ln(x) + 1 - 1$$

$$= \ln(x)$$

Thus we see that $x\ln(x) - x + c$ is indeed the antiderivative of $\ln(x)$. Showing that $\int \log_b(x)dx = x\log_b(x) - \frac{x}{\ln(b)} + c$ is similar.

Example 6.17

Find the antiderivative of $f(x) = 5\log_3(x)$.

Solution:

$$\int 5\log_3(x)\, dx = 5x\log_3(x) - \frac{5x}{\ln(3)} + c$$

There are two more integral rules we will present. These are the integrals of specific instances of the chain rule. We will present the general situation in the next chapter.

Indefinite Integrals Involving the Chain Rule:

Given a function $f(x)$ whose derivative is $f'(x)$, we have

$$\int f'(x)e^{f(x)}\, dx = e^{f(x)} + c$$

and

$$\int \frac{f'(x)}{f(x)}\, dx = \ln|f(x)| + c.$$

Example 6.18

Prove both of these rule by taking the derivative of the right-hand side.

Solution: For the first rule, we have $f(x)$ as the "inside" function and e^x as the "outside" function. The derivative of e^x is e^x and the derivative of $f(x)$ is $f'(x)$. Combining in the chain rule, we have

$$\left[e^{f(x)}\right]' = e^{f(x)} \cdot f'(x) = f'(x)e^{f(x)},$$

thus proving the first rule. For the second rule, we have $f(x)$ as the "inside" function and $\ln(x)$ as the "outside" function. The derivative of $\ln(x)$ is $\frac{1}{x}$ and the derivative of $f(x)$ is $f'(x)$. Combining in the chain rule, we have

$$\left[\ln|f(x)|\right]' = \frac{1}{f(x)} \cdot f'(x) = \frac{f'(x)}{f(x)},$$

thus proving the second rule.

Example 6.19

Find the indefinite integral of $f(x) = 4e^{4x}$.

Solution:

$$\int 4e^{4x}\, dx = e^{4x} + c$$

Example 6.20

Find the indefinite integral of $f(x) = \frac{4x^3+3x^2}{x^4+x^3}$.

Solution:

$$\int \frac{4x^3 + 3x^2}{x^4 + x^3} = \ln\left|x^4 + x^3\right| + c$$

Let us now briefly consider one simple application. Suppose we knew that a marginal revenue function was given by

$$R'(x) = -0.08x + 400.$$

We want to find the revenue function $R(x)$ associated with this marginal revenue function. In other words, we are given $R'(x)$ and are trying to find $R(x)$, which is the antiderivative of $R'(x)$. Using our rules, it is easy to see that

$$R(x) = \int -0.08x + 400 \, dx$$

$$= -0.08\frac{x^2}{2} + 400x + c$$

$$= -0.04x^2 + 400x + c.$$

But notice, the revenue function we have found has an unknown constant c in it. Without additional information, this is as far as we can go. Thus, we could not find the exact revenue function, only a general revenue function that includes the number c. But suppose we had some additional information, that when sales were zero we had no revenue. In other words, we have $R(0) = 0$. This is a very reasonable assumption to make regarding revenue in most situations. Thus, we have

$$R(0) = -0.04(0)^2 + 400(0) + c = 0,$$

which easily gives us $c = 0$. Thus, we can say that the revenue function is

$$R(x) = -0.04x^2 + 400x.$$

Example 6.21

Suppose the marginal revenue for a certain product is given by $MR = R'(x) = -0.025x + 2000$ and we know that the revenue is zero when sales are zero. Find the revenue function.

Solution: We first obtain a general revenue function by taking the antiderivative of the marginal revenue function,

$$R(x) = \int -0.025x + 2000 \; dx$$

$$= -0.025\frac{x^2}{2} + 2000x + c$$

$$= -0.0125x^2 + 2000x + c.$$

We now use the fact that $R(0) = 0$ to find c,

$$R(0) = -0.0125(0)^2 + 2000(0) + c = 0,$$

which gives us $c = 0$. Thus, the revenue function is

$$R(x) = -0.0125x^2 + 2000x.$$

Example 6.22

Suppose the marginal cost of producing x units of a product is given by $MC = C'(x) = 0.9x^2 + 0.5x$ and fixed costs are \$20,000. Find the cost function.

Solution. We first find the general cost function by finding the antiderivative of the marginal cost function,

$$C(x) = \int 0.9x^2 + 0.5x \; dx$$

$$= 0.9\frac{x^3}{3} + 0.5\frac{x^2}{2} + c$$

$$= 0.3x^3 + 0.25x^2 + c.$$

When we say that fixed costs are \$20,000, what we mean is that when production is zero, the costs are \$20,000. In other words, $C(0) = 20,000$. Thus, we have

$$C(0) = 0.3(0)^3 + 0.25(0)^2 + c = 20,000.$$

It is easy to see that $c = 20,000$. This gives a cost function of

$$C(x) = 0.3x^3 + 0.25x^2 + 20,000.$$

6.2 DEFINITION AND GRAPHICAL MEANING OF THE INTEGRAL

In this section, we will explore the geometrical meaning behind the integral and explain the notation in somewhat more detail. However, we caution you that our presentation will not be as rigorous as one you would receive in a calculus class meant for science, technology, engineering or mathematics students; we are more interested in applications than in mathematical rigor. Nor will it be immediately obvious how this relates to the preceding section, though it does.

We will begin by considering an example. Suppose we had the function $f(x) = 3x$ and were interested in finding the area underneath the graph between 0 and 20. In other words, we are interested in finding the area of the region bounded by $f(x) = 3x$, the interval $[0, 20]$ on the x-axis, and the vertical line $x = 20$, as shown below.

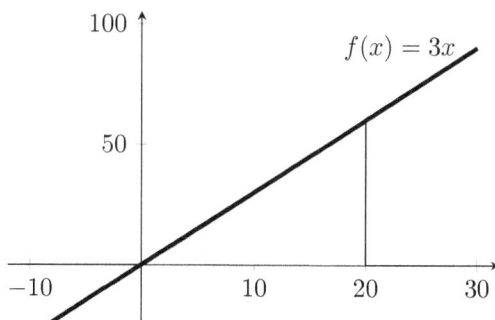

This is of course an area that we could readily find using grade school geometry, the area of the shaded triangle is given by

$$\text{Area} = \frac{1}{2}\text{base} \cdot \text{height}.$$

The base of the triangle is obviously 20 and the height of this triangle is given by $f(20) = 60$,

$$\text{Area} = \frac{1}{2} \cdot 20 \cdot 60 = 600.$$

We have, of course, omitted units in this calculation. However, in general the graphs of functions $f(x)$ are much more complicated than this and we can not use simple geometry. We need a more general technique to find the area under curves. However, we will continue to use this simple example to clearly illustrate the technique we will use.

Consider the interval $[0, 20]$, which is the interval over which we are wanting to find the area of under the graph of $f(x)$. Suppose we divide this interval into $n = 2$ equal subintervals. The width of the interval is 20 so each subinterval has a width of $\Delta x = \frac{20}{2} = 10$. Thus, the first subinterval is $[0, 10]$, the second subinterval is $[10, 20]$, and Δx is used to represent the widths of the subintervals.

We then select the midpoint of each interval, in this cases $x = 5$ is the midpoint of $[0, 10]$ and $x = 15$ is the midpoint of $[10, 20]$. We can then estimate the area under $f(x) = 3x$ by using two rectangles, both of width 10, the first with height $f(5) = 15$ and the second with height $f(15) = 45$.

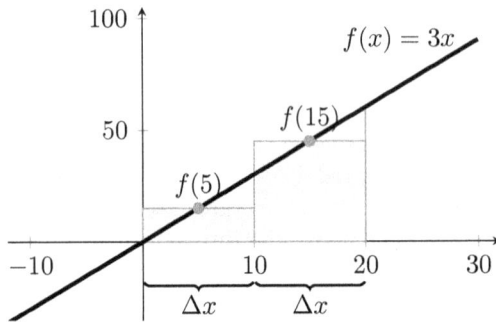

$$\text{Area} \approx f(5) \cdot \Delta x + f(15) \cdot \Delta x$$
$$= 15 \cdot 10 + 45 \cdot 10$$
$$= 600.$$

In this case our approximation turns out to be the exact number because we have a linear function, but for more complicated functions this will not happen and our estimate would be off. In general, the smaller the width of the interval Δx the more accurate the approximation generated by this procedure would be. Suppose instead of $n = 2$ we chose $n = 4$ intervals. This gives us

$$\Delta x = \frac{20}{4} = 5.$$

Splitting the interval $[0, 20]$ into $n = 4$ subintervals gives $[0, 5]$, $[5, 10]$, $[10, 15]$, and $[15, 20]$. Again, we choose the midpoint of each of these subintervals to calculate the heights of our rectangles $f(2.5) = 7.5$, $f(7.5) = 22.7$, $f(12.5) = 37.5$, and $f(17.5) = 52.5$.

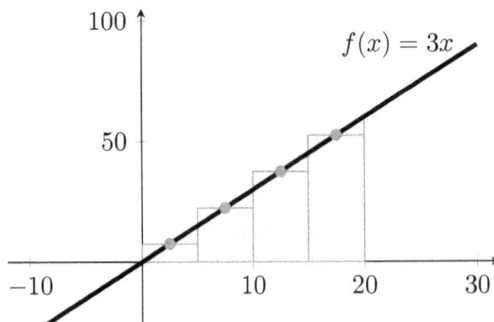

The area is approximated by the sum of these four rectangles

$$\text{Area} \approx f(2.5) \cdot \Delta x + f(7.5) \cdot \Delta x + f(12.5) \cdot \Delta x + f(17.5) \cdot \Delta x.$$

In general, this would be a more accurate estimate, but would still not give an exact answer. Again, we could improve this estimate by increasing the number of rectangles. The picture corresponding to $n = 8$ is shown below.

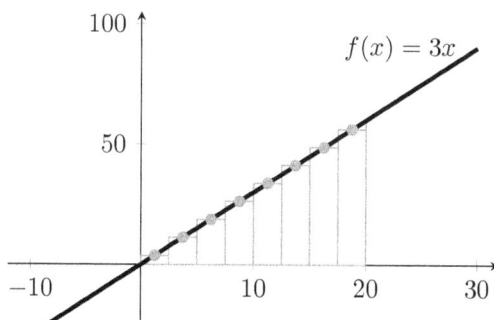

The area under the curve would be approximated by the sum

$$\sum_{i=1}^{8} f(x_i)\Delta x = f(x_1)\Delta x + f(x_2)\Delta x + f(x_3)\Delta x + f(x_4)\Delta x$$

$$+ f(x_5)\Delta x + f(x_6)\Delta x + f(x_7)\Delta x + f(x_8)\Delta x,$$

where x_i represents the midpoint of the ith subinterval. Here we are, of course, using summation notation. This sum is called a **Riemann Sum**

Clearly, the more rectangles we have, the more accurate our estimate of the area under the curve of $f(x)$. If we allowed $n \to \infty$ then $\Delta x \to 0$ and in the limit our answer would become exact. This leads us to the actual definition of the definite integral.

Definition of the Definite Integral

Suppose we are given an interval $[a, b]$, where $a < b$, and a function $f(x)$ where $f(x) > 0$ for all x in the interval $[a, b]$, then the definite integral of $f(x)$ from a to b is given by

$$\int_a^b f(x)dx = \lim_{n \to \infty} \sum_{i=1}^{n} f(x_i)\Delta x,$$

where a and b are the limits of integration, with a being the **lower bound** and b being the **upper bound**. The definite integral $\int_a^b f(x)dx$ gives the area of the region bounded by $x = a$, $x = b$, the x-axis, and $f(x)$.

The $\int$ sign is simply an elongated letter "S" which represents the word "sum." We can think of $\int$ as simply what happens to the $\sum$ as we let $n \to \infty$. And clearly the dx is what happens to the Δx as $\Delta x \to 0$.

Since we used the integral notation in the last section when discussing antiderivatives, you will not be surprised to find that antiderivatives are intimately related to definite integrals, though we will not discuss the actual relation until the next section.

Example 6.23

Sketch the function $f(x) = -x^2 + 4x + 4$ and shade the area under $f(x)$ for $1 \leq x \leq 4$. Indicate this area using a definite integral.

Solution: The area of the indicated region is given by $\int_1^4 (-x^2 + 4x + 4) dx$, shown below.

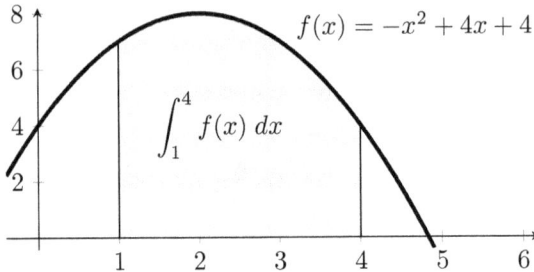

Example 6.24

Estimate the area under the function $f(x) = -x^2 + 4x + 4$ for $1 \leq x \leq 4$ using $n = 6$ rectangles.

Solution: Since we want $n = 6$ rectangles, the width of each rectangle will be $\Delta x = \frac{4-1}{6} = \frac{1}{2}$ units wide. Thus, the interval $[1, 4]$ is divided into $n = 6$ subintervals of equal width $\frac{1}{2}$. The x_i are the midpoints of each of these subintervals. To make things easier, we start with a table to collect our values.

x_i	1.25	1.75	2.25	2.75	3.25	3.75
$f(x_i)$	7.4375	7.9375	7.9375	7.4375	6.4375	4.9375

We now compute the Riemann sum using these values. To make the computation a little easier, we can notice that each term is multiplied by $\Delta x = \frac{1}{2}$, which can be done collectively after the terms are added. Algebraically, this is equivalent to factoring out the Δx.

$$\sum_{i=1}^{6} f(x_i)\Delta x = \Big[f(x_1) + f(x_2) + f(x_3) + f(x_4) + f(x_5) + f(x_6)\Big]\Delta x$$

$$= \Big[7.4375 + 7.9375 + 7.9375 + 7.4375 + 6.4375 + 4.9375\Big]\frac{1}{2}$$

$$= 21.0625.$$

This value is quite close to the exact answer, which is 21. (We will discuss how to find the exact answer in the next section). The rectangles are illustrated below.

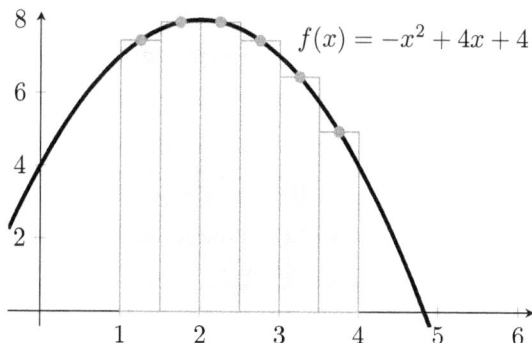

A few words of warning are necessary here. In our Riemann sums, the x_i denotes the midpoint of the ith subinterval. However, in general, the x_i need not be the midpoint. Other popular choices for x_i are the left-endpoint of the interval or the right-endpoint of the interval. In reality, any point in the interval will suffice. However, in this book we will keep things simple and always assume x_i is the midpoint.

Another question also arises. In the definition of the definite integral, we stated that $f(x) > 0$ on the interval we are integrating over. But is that strictly necessary? Let us consider the above example, only with the function reflected across the x-axis. The following example proceeds exactly as before, only with a few sign changes.

Example 6.25

Estimate the area under the function $f(x) = x^2 - 4x - 4$ for $1 \leq x \leq 4$ using $n = 6$ rectangles.

Solution: Again, since we want $n = 6$ rectangles, the width of each rectangle will be $\Delta x = \frac{4-1}{6} = \frac{1}{2}$ units wide. Thus, the interval $[1, 4]$ is divided into $n = 6$ subintervals of equal width $\frac{1}{2}$. The x_i are the midpoints of each of these subintervals. Again, to make things easier, we start with a table to collect our values.

x_i	1.25	1.75	2.25	2.75	3.25	3.75
$f(x_i)$	-7.4375	-7.9375	-7.9375	-7.4375	-6.4375	-4.9375

But here we notice that the $f(x_i)$ values are negative instead of positive. This should not surprise us. We simply continue with computing the

Riemann sum using these negative values.

$$\sum_{i=1}^{6} f(x_i)\Delta x = \left[f(x_1) + f(x_2) + f(x_3) + f(x_4) + f(x_5) + f(x_6) \right]\Delta x$$

$$= \left[-7.4375 - 7.9375 - 7.9375 - 7.4375 - 6.4375 - 4.9375 \right]\frac{1}{2}$$

$$= -21.0625.$$

This is exactly the numerical answer that we got before, only with a negative sign. In order to find the actual area, as you understand it, simply take the absolute value of the number you have obtained. Thus, the area would be $|-21.0625| = 21.0625$. The rectangles are illustrated below.

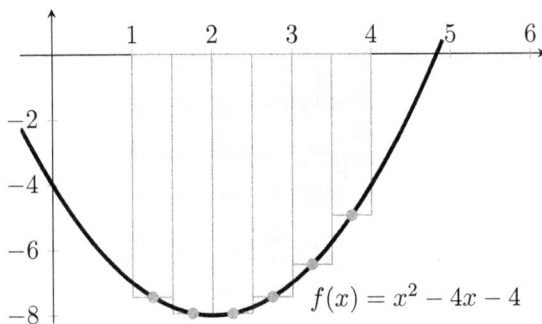

$$f(x) = x^2 - 4x - 4$$

While the idea of a negative area may come as a surprise to you, the fact of the matter is that areas do indeed come with a sign (positive or negative) attached. This is beyond the scope of this book, but it ultimately depends upon what is called the orientation of the coordinate system. But this quirk does explain why we made the assumption that $f(x) > 0$. In practical terms, this means we have to handle areas under the x-axis separately from areas above the x-axis. We will go into this in greater detail in the next section.

Example 6.26

Estimate the area under the curve $f(x) = 2\ln(x)$ for $3 \le x \le 7$ using $n = 4$ rectangles.

Solution: The width of each rectangle is given by $\Delta x = \frac{7-3}{4} = 1$. We divide the interval $[3, 7]$ into 4 subintervals and then choose the midpoint of each subinterval to find $f(x_i)$. We collect our values in a table.

x_i	3.5	4.5	5.5	6.5
$f(x_i)$	2.5055	3.0082	3.4095	3.7436

The Riemann sum is

$$\sum_{i=1}^{4} f(x_i)\Delta x = \left[f(x_1) + f(x_2) + f(x_3) + f(x_4)\right]\Delta x$$

$$= \left[2.5055 + 3.0082 + 3.4095 + 3.7436\right] \cdot (1)$$

$$= 12.6668.$$

The exact answer, to four decimal places, is 12.6511, so our approximation with four rectangles is reasonably accurate.

6.3 FUNDAMENTAL THEOREM OF CALCULUS

In section one of this chapter, we discussed the antiderivative of a function $f(x)$. The function $F(x)$ is called an antiderivative of $f(x)$ if $F'(x) = f(x)$. We also said that antiderivatives were called indefinite integrals, and introduced the integral notation, which is what we used to write out the various antiderivative rules;

$$\int f(x)dx = F(x).$$

In section two of this chapter, we discussed how areas under the graph of $f(x)$ and above an interval $[a, b]$ were defined as definite integrals by

$$\int_a^b f(x)dx = \lim_{n\to\infty} \sum_{i=1}^{n} f(x_i)\Delta x.$$

Now we present the theorem that actually relates the perspectives of sections one and two. This theorem is called the **Fundamental Theorem of Calculus**. This theorem is both deep and extremely important, and it explains why we use the integral notation to write the antiderivative rules. We will present this theorem and explain what it means and how it is used in this section, but we will not try to prove it. That is beyond the scope of this book.

The Fundamental Theorem of Calculus

Suppose that $f(x)$ is a continuous function on the closed interval $[a, b]$ and that $F(x)$ is an antiderivative of $f(x)$ on $[a, b]$. Then

$$\int_a^b f(x)dx = F(b) - F(a).$$

In other words, the area under $f(x)$ on the interval $[a, b]$ is given by the antiderivative evaluated at b minus the antiderivative evaluated at a.

This is a deeply surprising result; the area under the graph of a function f is intimately related to the antiderivative of the function f. This powerful result also is incredibly useful in a wide range of applications. But before we turn to some examples let us mention some common notations. The right-hand side of the Fundamental Theorem of Calculus is often written in one of these two ways,

$$F(b) - F(a) = \left[F(x) \right]_a^b,$$

or as

$$F(b) - F(a) = F(x) \Big|_a^b.$$

We now turn to a few examples that build on some examples from the last section.

Example 6.27

Find the area under the graph of $f(x) = -x^2 + 4x + 4$ over the interval $[1, 4]$.

Solution: In Example 6.23 we learned that the area of the indicated region is given by the definite integral $\int_1^4 (-x^2 + 4x + 4) dx$, which was then defined as the limit of the Riemann sum as $n \to \infty$. In Example 6.24 uses a Riemann sum with $n = 6$ to approximate this area and found it to be 21.0625. In that example we said the exact answer was given by 21. Here we find that exact answer using the Fundamental Theorem of Calculus.

$$
\begin{aligned}
\int_1^4 (-x^2 + 4x + 4) \, dx &= \left[-\frac{x^3}{3} + 2x^2 + 4x + c \right]_1^4 \\
&= \left(-\frac{4^3}{3} + 2 \cdot 4^2 + 4 \cdot 4 + c \right) \\
&\quad - \left(-\frac{1^3}{3} + 2 \cdot 1^2 + 4 \cdot 1 + c \right) \\
&= \left(-\frac{64}{3} + 32 + 16 + c \right) - \left(-\frac{1}{3} + 2 + 4 + c \right) \\
&= -\frac{64}{3} + 32 + 16 + c + \frac{1}{3} - 2 - 4 - c \\
&= 21.
\end{aligned}
$$

Notice what happens to the constants c; they simply cancel each other out. Thus, when evaluating definite integrals the standard practice is to simply not to bother with writing the constant c for the antiderivative.

The graph is shown below. The regions whose area we wanted to find is shaded.

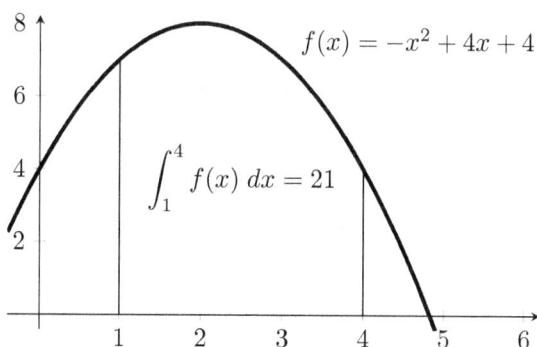

$$\int_1^4 f(x)\, dx = 21$$

$$f(x) = -x^2 + 4x + 4$$

This is a typical use of the Fundamental Theorem of Calculus. In this example the function $f(x)$ was positive in the interval over which we were integrating. But what happens if the function is negative over the interval of integration instead of positive? How does this affect the answer, if at all? We explore this in the next example.

Example 6.28

Find area under the graph of $f(x) = x^2 - 4x - 4$ over the interval $[1, 4]$.

Solution: In Example 6.25 we used a Riemann sum with $n = 6$ rectangles to estimate this area and found it to be -21.0625. Here we will compute the exact number. This is very similar to Example 6.27, only with all the signs switched.

$$\int_1^4 \left(x^2 - 4x - 4\right) dx = \left[\frac{x^3}{3} - 2x^2 - 4x\right]_1^4$$

$$= \left(\frac{4^3}{3} - 2 \cdot 4^2 - 4 \cdot 4\right) - \left(\frac{1^3}{3} - 2 \cdot 1^2 - 4 \cdot 1\right)$$

$$= \left(\frac{64}{3} - 32 - 16\right) - \left(\frac{1}{3} - 2 - 4\right)$$

$$= -21.$$

Thus, the Fundamental Theorem of Calculus gives us a negative area just like the estimate in Example 6.25 did. The graph is shown below. The region whose area we wanted to find is shaded.

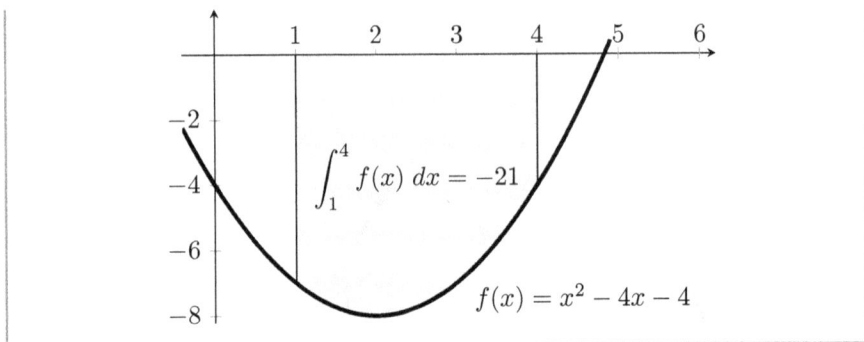

$$\int_1^4 f(x)\,dx = -21$$

$$f(x) = x^2 - 4x - 4$$

The last two examples illustrate one issue with signs that can arise when taking derivatives. Another issue with signs is explored in the next few examples.

Example 6.29

Find the area under the graph of $f(x) = 4x^3$ over the interval $[1, 2]$.

Solution: Using the Fundamental Theorem of Calculus, we find the following definite integral:

$$\int_1^2 4x^3\,dx = \left[x^4\right]_1^2 = (2^4) - (1^4) = 15.$$

This is shown below:

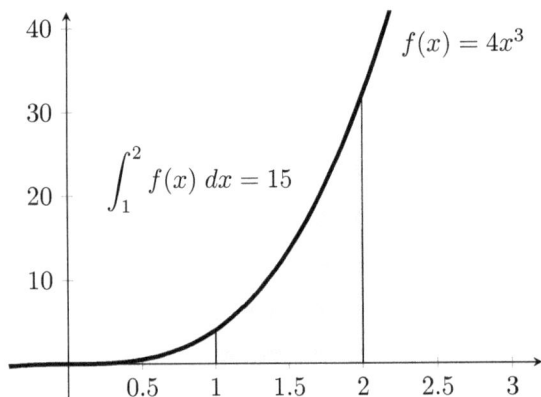

$$f(x) = 4x^3$$

$$\int_1^2 f(x)\,dx = 15$$

What happens when we switch the limits of integration?

Example 6.30

Find the following definite integral using the Fundamental Theorem of Calculus and explain why the answer differs from Example 6.29.

$$\int_2^1 4x^3\, dx$$

Solution: Notice, in this definite integral our limits of integration are the reverse of those from Example 6.29; our lower and upper limits of integration have been switched. One way to think about this is that we integrate from right to left instead of from left to right. However, when performing the computation, our application of the Fundamental Theorem of Calculus remains exactly the same,

$$\int_2^1 4x^3\, dx = \left[x^4\right]_2^1 = \left(1^4\right) - \left(2^4\right) = -15.$$

Reversing the limits of integration changed the sign of the area, though the shaded area looks exactly the same. This is shown below.

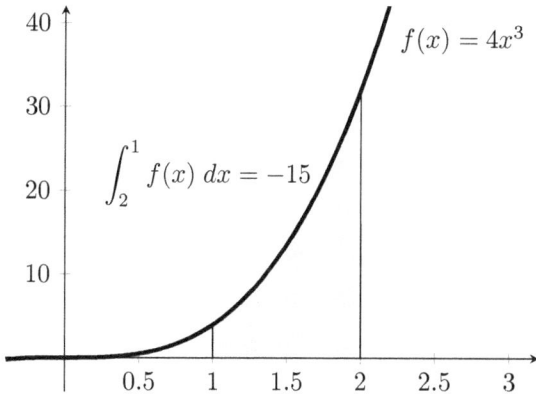

In summary, we have the following situations:

1. If $f(x) > 0$ on $[a, b]$ (where $a < b$) then $\int_a^b f(x)\,dx$ is positive.

2. If $f(x) > 0$ on $[a, b]$ (where $a < b$) then $\int_b^a f(x)\,dx$ is negative.

3. If $f(x) < 0$ on $[a, b]$ (where $a < b$) then $\int_a^b f(x)\,dx$ is negative.

4. If $f(x) < 0$ on $[a, b]$ (where $a < b$) then $\int_b^a f(x)\,dx$ is positive.

Example 6.31

Given $f(x) = 2x$ find the following:

(a) $\int_1^2 f(x)dx$ (c) $\int_{-2}^{-1} f(x)dx$ (e) $\int_{-1}^1 f(x)dx$

(b) $\int_2^1 f(x)dx$ (d) $\int_{-1}^{-2} f(x)dx$ (f) $\int_1^1 f(x)dx$

Solution: Using the Fundamental Theorem of Calculus, we get

(a) $\int_1^2 f(x)dx = \left[x^2\right]_1^2 = 2^2 - 1^2 = 3$

(b) $\int_2^1 f(x)dx = \left[x^2\right]_2^1 = 1^2 - 2^2 = -3$

(c) $\int_{-2}^{-1} f(x)dx = \left[x^2\right]_{-2}^{-1} = (-1)^2 - (-2)^2 = -3$

(d) $\int_{-1}^{-2} f(x)dx = \left[x^2\right]_{-1}^{-2} = (-2)^2 - (-1)^2 = 3$

(e) $\int_{-1}^1 f(x)dx = \left[x^2\right]_{-1}^1 = 1^2 - (-1)^2 = 0$

(f) $\int_1^1 f(x)dx = \left[x^2\right]_1^1 = 1^2 - 1^2 = 0$

Notice, in (e) the positive and negative areas cancel each other out and in (f) if the lower and upper limits of integration are the same, the area is zero.

Example 6.32

Find the area under the graph of $f(x) = 2^x$ between $x = -1$ and $x = 2$.

Solution:
$$\int_{-1}^2 2^x \, dx = \left[\frac{2^x}{\ln(2)}\right]_{-1}^2$$
$$= \left(\frac{2^2}{\ln(2)}\right) - \left(\frac{2^{-1}}{\ln(2)}\right)$$
$$= \frac{4}{\ln(2)} - \frac{1}{2\ln(2)}$$
$$= \frac{7}{2\ln(2)}$$

The area found is shaded below.

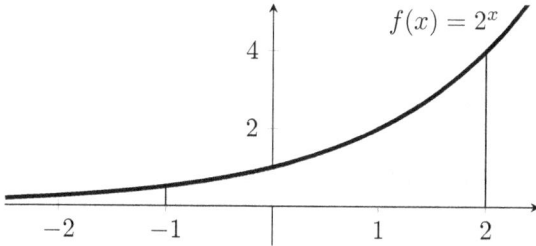

6.4 AREAS BETWEEN CURVES

One of the primary uses for the Fundamental Theorem of Calculus is to find the area between two curves. Suppose we wanted to find the area between the curve $f(x) = -x^2 + 4x + 4$ and the horizontal line given by $g(x) = 3$ between $x = 1$ and $x = 4$, shown below.

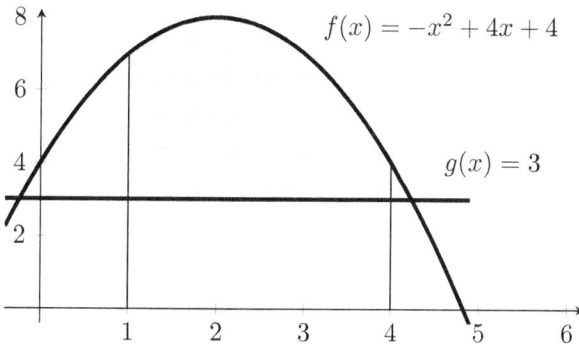

Notice that on the interval $[1,4]$ we have $f(x) \geq g(x)$. One way to think about doing this is to find the area under the curve $f(x) = -x^2 + 4x + 4$, then find the area under the curve $g(x) = 3$, and then subtract this from the area we just found to give us the area between the two curves;

$$\text{Area between } f(x) \text{ and } g(x) = \int_1^4 f(x)\,dx - \int_1^4 g(x)\,dx.$$

Graphically, this looks like:

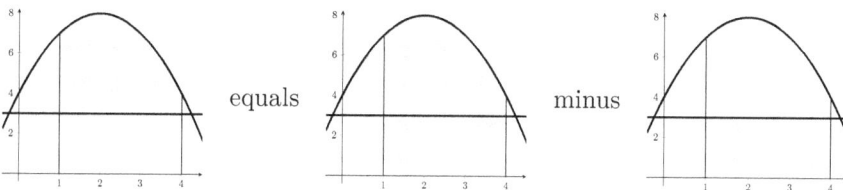

 equals minus

Using the integral rule for sums and differences, we have

$$\int_1^4 f(x)\, dx - \int_1^4 g(x)\, dx = \int_1^4 f(x) - g(x)\, dx.$$

In other words,

$$\text{Area between } f(x) \text{ and } g(x) = \int_1^4 f(x) - g(x)\, dx$$

This example illustrates the general formula.

The Area Between Two Curves:

Suppose that $f(x) \geq g(x)$ on the interval $[a, b]$. Then the area between $f(x)$ and $g(x)$ from $x = a$ to $x = b$ is given by

$$\text{Area} = \int_a^b f(x) - g(x)\, dx.$$

The curve $f(x)$ is sometimes called the **upper curve** and $g(x)$ is sometimes called the **lower curve**.

We now consider a few examples.

Example 6.33

Find the area between $f(x) = -x^2 + 4x + 4$ and $g(x) = 3$ on the interval $[1, 4]$.

Solution:
$$\text{Area} = \int_1^4 \left(-x^2 + 4x + 4 \right) - 3 \, dx$$

$$= \int_1^4 -x^2 + 4x + 1 \, dx$$

$$= \left[-\frac{x^3}{3} + 2x^2 + x \right]_1^4$$

$$= \left(-\frac{64}{3} + 32 + 4 \right) - \left(-\frac{1}{3} + 2 + 1 \right)$$

$$= 12$$

Example 6.34

Find the area enclosed by the graphs of $f(x) = x^2 - 6x + 9$ and $g(x) = x - 1$.

Solution: In order to determine the x-values needed for the limits of integration, we need to find where the graphs of $f(x)$ and $g(x)$ intersect. This occurs when the two graphs are equal to each other.

$$f(x) = g(x) \implies x^2 - 6x + 9 = x - 1$$
$$\implies x^2 - 7x + 10 = 0$$
$$\implies (x - 5)(x - 2) = 0$$
$$\implies x = 5 \text{ or } x = 2.$$

The area between $f(x)$ and $g(x)$ is shown below.

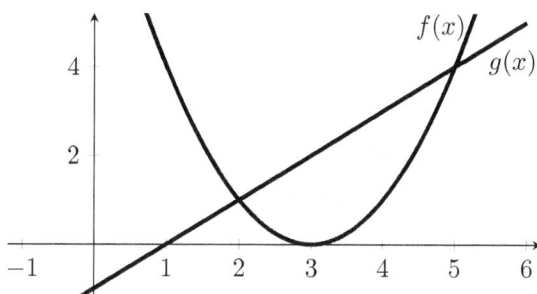

With this we can then set up the necessary definite integral to find the area between the two curves. Be careful though, here the upper curve is actually $g(x)$ and the lower curve is $f(x)$. Thus, we have

$$\int_2^5 g(x) - f(x)\, dx = \int_2^5 (x - 1) - (x^2 - 6x + 9)\, dx$$
$$= \int_2^5 -x^2 + 7x - 10\, dx$$
$$= \left[-\frac{x^3}{3} + \frac{7x^2}{2} - 10x \right]_2^5$$
$$= \left(-\frac{125}{3} + \frac{175}{2} - 50 \right) - \left(-\frac{8}{3} + \frac{28}{2} - 20 \right)$$
$$= \frac{9}{2}.$$

In the last example we had $g(x) \geq f(x) \geq 0$ over the interval $[2, 5]$, which we were integrating over. It turns out that it is not necessary for the two curves to be greater than 0 over the interval of integration. A careful analysis of the Riemann sums associated with the integral for areas between curves will show that the signs work out and we indeed arrive at the correct area. It is sufficient that $f(x) \geq g(x)$. We will not go through that analysis here, but it is easy enough to do if you are interested.

Example 6.35

Find the area between the curves $f(x) = x^2 + 3$ and $g(x) = 2x - 4$, from $x = 0$ to $x = 3$.

Solution: A portion of this region lies below the x-axis. in the last section we learned that regions below the x-axis result in negative areas. However, for integral for the area between two curves, we do not need to worry about this; we can simply carry out the computation.

$$\int_0^3 f(x) - g(x)\, dx = \int_0^3 \left(x^2 + 3\right) - \left(2x - 4\right) dx$$

$$= \int_0^3 x^2 - 2x + 7\, dx$$

$$= \left[\frac{x^3}{3} - x^2 + 7x\right]_0^3$$

$$= \left(\frac{3^3}{3} - 3^2 + 7 \cdot 3\right) - \left(\frac{0^3}{3} - 0^2 + 0\right)$$

$$= 21.$$

The region is shown below.

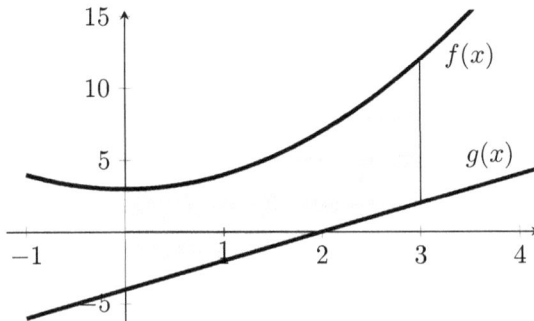

6.5 ANTIDERIVATIVES AND THE INTEGRAL IN BUSINESS APPLICATIONS

Now that we have discussed definite integrals as giving the area of a region under a curve and the use of the Fundamental Theorem of Calculus in evaluating definite integrals, we turn our attention to the meaning of the area under curves in various business problems. In many applications, including many business applications, the integrand function, that is, the function that is being integrated, is a function that describes rates of change of some quantity. The definite integral then gives the total change in that quantity. We will see this application of integration for the cost and revenue functions.

Cost Function

Suppose we have the cost function $C(x) = 2x$; that is, the total cost of producing x items is $2x$. For example, if no items are produced, the cost of production is zero, if ten items are produced the cost of production is 20. In general, the unit for cost is dollars, or some other currency.

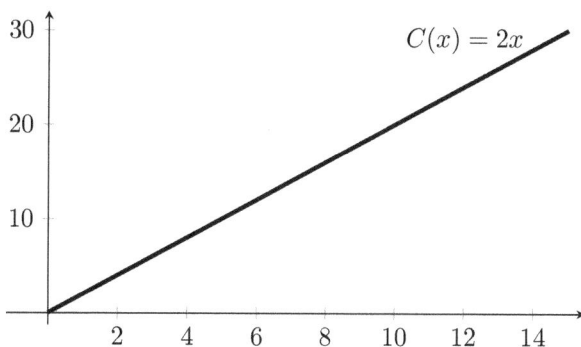

Now consider the marginal cost function, which is simply the derivative of the cost function, $C'(x) = 2$. We know derivatives are instantaneous rates of change, which are obtained by taking the limit of the difference quotient.

$$C'(x) = \lim_{h \to 0} \frac{C(x+h) - C(x)}{(x+h) - x} = \lim_{h \to 0} \frac{C(x+h) - C(x)}{h},$$

which has units of dollars per item produced. Suppose we graphed the marginal costs function. In this particular case, we would simply have a horizontal line at 2. What would units be for the y-axis and the x-axis? In this example the y-axis units would be $\frac{\text{dollars}}{\text{item}}$ and the units for the x-axis would be items.

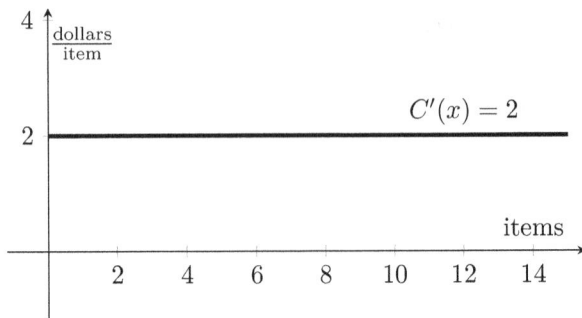

What exactly would the area under this curve represent? Suppose we wanted to find the area under $C'(x)$ between $x = 0$ and $x = 10$, as shown.

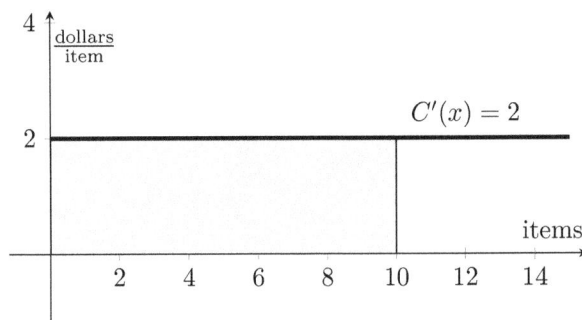

This is, of course, simply a rectangle so we can get the area by multiplying the length by the width of the rectangle. Including units gives us

$$\text{length} \cdot \text{width} = \left(10 \text{ items}\right) \cdot \left(2 \frac{\text{dollars}}{\text{item}}\right) = 20 \text{ dollars}.$$

Notice that this is exactly

$$\int_0^{10} 2 \, dx = 20,$$

This area represents the total cost of producing the first ten items. In other words, this area exactly the same as $C(10) = 2 \cdot 10 = 20$ dollars. Since $C(0) = 2 \cdot 0 = 0$ dollars, we have

$$\int_0^{10} C'(x) \, dx = C(10) - C(0),$$

which is what we would expect from the Fundamental Theorem of Calculus. Even the units work out perfectly and our answer is in dollars. In other words, integrating the marginal cost function $MC = C'(x)$ from $x = a$ to $x = b$ gives the cost of producing items a through b;

$$\int_a^b C'(x) \, dx = C(b) - C(a).$$

Example 6.36

Given the marginal cost function $C'(x) = 2$, find the cost of producing the 11th through the 15th items.

Solution: We need to include the 11th item in our integration, hence we want to integrate from $x = 10$ to $x = 15$,

$$\int_{10}^{15} 2\,dx = \Big[2x\Big]_{10}^{15}$$
$$= \Big(2 \cdot 15\Big) - \Big(2 \cdot 10\Big)$$
$$= 10.$$

Thus, producing items 10 through 15 costs 10 dollars.

It is clear that what holds true for cost functions also holds true for revenue and profit functions. Thus, this fairly simple example illustrates one of the fundamental uses for integration in business applications.

Example 6.37

Suppose you had the marginal cost function $C'(x) = 0.009x^2 - 0.2x + 20$. Find the following:

1. The cost of producing the first 20 items.
2. The cost of producing the 51st through the 100th items.
3. The cost of producing the 25th item.

Solution:

1. The cost of producing the first 20 items means we need to find the area under the marginal cost function from $x = 0$ to $x = 20$.

$$\int_{0}^{20} 0.009x^2 - 0.2x + 20\,dx = \left[\frac{0.009x^3}{3} - \frac{0.2x^2}{2} + 20x\right]_{0}^{20}$$
$$= \Big(0.003(20)^3 - 0.1(20)^2 + 20(20)\Big)$$
$$- \Big(0.003(0)^3 - 0.1(0)^2 + 20(0)\Big)$$
$$= 384.$$

Thus, producing the first 20 items costs 384 dollars.

2. The cost of producing the 51st through the 100th items means we need to find the area under the marginal costs function from $x = 50$ to $x = 100$.

$$\int_{50}^{100} 0.009x^2 - 0.2x + 20 \; dx = \left[\frac{0.009x^3}{3} - \frac{0.2x^2}{2} + 20x \right]_{50}^{100}$$

$$= \left(0.003(100)^3 - 0.1(100)^2 + 20(100) \right)$$

$$- \left(0.003(50)^3 - 0.1(50)^2 + 20(50) \right)$$

$$= 2875.$$

3. The cost of producing exactly the 25th item means we want to find the area under the curve between $x = 24$ and $x = 25$.

$$\int_{24}^{25} 0.009x^2 - 0.2x + 20 \; dx = \left[\frac{0.009x^3}{3} - \frac{0.2x^2}{2} + 20x \right]_{24}^{25}$$

$$= \left(0.003(25)^3 - 0.1(25)^2 + 20(25) \right)$$

$$- \left(0.003(24)^3 - 0.1(24)^2 + 20(24) \right)$$

$$= 20.50.$$

As usual, with money we generally round to two decimal places.

Let us explore a slight variation of the last problem. Suppose we were given the cost function

$$C(x) = 0.003x^3 - 0.1x^2 + 20x + 500$$

whose marginal cost function is simply

$$C'(x) = 0.009x^2 - 0.2x + 20,$$

the same marginal cost function as in the last example. Notice that $C(0) = 500$, which represents the fixed costs. Now suppose we asked, what is the cost of producing twenty items? To answer that we would evaluate the cost function at 20,

$$C(20) = 0.003(20)^3 - 0.1(20)^2 + 20(20) + 500 = 884.$$

Thus, it costs \$884 to produce 20 items. Notice how different this is from the first answer of \$384 that we obtained in the last example. It is \$500 more. This is, of course, the fixed costs. So we need to be very careful here! Asking "what is the cost of producing the first 20 items?" is a subtly different question than asking "what is the total cost of producing 20 items?" In answering the first question we do not include the fixed costs, but in answering the second question we do.

Revenue Function

Just as with cost functions, revenue functions in business can be analyzed using antiderivatives and integrals. The total revenue, R, is a function of the number of units sold, x. The marginal revenue function, MR, represents the additional revenue generated by selling one more unit, and it is the derivative of the total revenue function, $R(x)$.

$$MR = \frac{dR}{dx} = R'(x)$$

The total revenue function $R(x)$ can be determined by integrating the marginal revenue function MR. If we assume that the revenue is zero when nothing is sold, this initial condition helps us determine the constant of integration.

$$R(x) = \int MR\, dx$$

The average revenue, which is the revenue per unit, can be found by dividing the total revenue by the number of units, x. This is equivalent to the price per unit, and hence, the average revenue curve coincides with the price-demand function. To see this recall

$$R(x) = \text{ price } \cdot \text{ quantity } = \frac{R(x)}{x} \cdot x,$$

which gives us the price-demand function $p = \frac{R(x)}{x}$.

Example 6.38

Given the marginal revenue function $MR = 3 - 2x + 9x^2$, find p.

Solution: To find the revenue function $R(x)$ corresponding to the given marginal revenue MR, we need to take the antiderivative of MR with respect to x:

$$R(x) = \int (3 - 2x + 9x^2)\, dx = 3x - x^2 + 3x^3 + c.$$

Using the initial condition $R(0) = 0$, we have $0 = 3(0) - 0^2 + 3(0)^3 + c$, which easily give us $c = 0$. Therefore, the total revenue function is

$$R(x) = 3x - x^2 + 3x^3.$$

The demand function, which is the same as the average revenue, is

$$p = \frac{R(x)}{x} = 3 - x + 3x^2.$$

Furthermore, understanding the relationship between total, average, and marginal revenue is crucial in business. For instance, the point at which marginal revenue equals zero is significant because it typically corresponds to the maximum total revenue. This concept is crucial in determining the most profitable level of production.

Example 6.39

Consider that the marginal revenue function is given by $MR = 4 - \frac{1}{2}x$. Determine the total revenue function and identify the quantity that maximizes the total revenue.

Solution: To find the revenue function $R(x)$ corresponding to the given marginal revenue MR, we need to take the antiderivative of $MR(x)$ with respect to x;

$$R(x) = \int MR\,dx$$

$$= \int \left(4 - \frac{1}{2}x\right)\,dx$$

$$= 4x - \frac{1}{4}x^2 + c.$$

Using $R(0) = 0$, we find $c = 0$. So, the total revenue function is $R(x) = 4x - \frac{1}{4}x^2$. To find the quantity that maximizes total revenue, we solve $MR = 0$;

$$4 - \frac{1}{2}x = 0 \implies x = 8.$$

Therefore, selling 8 units maximizes the total revenue.

Example 6.40

How does a business calculate its revenue if it sells a product at a price of $15 per unit?

Solution: The revenue function $R(x)$ can be represented as the product of the price per unit and the number of units sold, x. Thus, $R(x) = 15x$. For instance, selling 10 units would yield a revenue of $R(10) = 15 \times 10 = 150$ dollars.

The marginal revenue function, $R'(x)$, represents the additional revenue for each additional unit sold. In other words, $R'(x) = 15$ indicates a constant increase in revenue for each additional unit sold.

> **Example 6.41**
>
> How is the revenue function expressed and interpreted when it is more complex, such as $R(x) = 7x - 0.5x^2 + 0.05x^3$?
>
> Solution: For a complex revenue function like $R(x) = 7x - 0.5x^2 + 0.05x^3$, the marginal revenue, $R'(x)$, is obtained by differentiating $R(x)$, resulting in $R'(x) = 7 - x + 0.15x^2$. This revenue function indicates a non-linear relationship where the revenue initially increases with each unit sold but eventually decreases, a common scenario in markets experiencing saturation or price competition. As we have seen in Chapter 2, $R'(x)$ is a quadratic equation and has a critical point or a turning point at $x_0 = \frac{10}{3}$. This behavior suggests that there is an optimal number of units to sell beyond which the additional revenue starts to decrease.

Suppose a business sells a product at a price of $P(t)$ per unit, and the quantity sold over time is represented by $Q(t)$. To calculate the total revenue over a specific period from t_1 to t_2, you can use the integral:

$$TR = \int_{t_1}^{t_2} P(t) \cdot Q(t)\, dt$$

This formula integrates the product of the price and quantity over the given time interval to find the accumulated revenue.

> **Example 6.42**
>
> A company sells toys at a price of $P(t) = 10 - 0.5t$ dollars per toy, and the quantity sold is given by $Q(t) = 10t - 2t^2$ toys over the time interval from $t = 0$ to $t = 4$ hours. Calculate the total revenue during this period.
>
> Solution: We use the total revenue formula over the specific period,
>
> $$\begin{aligned} TR &= \int_0^4 (10 - 0.5t) \cdot (10t - 2t^2)\, dt \\ &= \int_0^4 (t^3 - 25t^2 + 100t)\, dt \\ &= \left[\frac{t^4}{4} - \frac{25t^3}{3} + 50t^2 \right]_0^4 \\ &= \left(\frac{4^4}{4} - \frac{25 \cdot 4^3}{3} + 50 \cdot 4^2 \right) - \left(\frac{0^4}{4} - \frac{25 \cdot 0^3}{3} + 50 \cdot 0^2 \right) \\ &= \frac{992}{3} = 330.67 \text{ dollars.} \end{aligned}$$

In business and economics, integration plays a crucial role in calculating profit functions as well. Given the marginal cost $C'(x)$ and marginal revenue $R'(x)$ functions, you can determine the profit function $P(x)$ as follows:

$$C(x) = \int C'(x)\, dx$$

$$R(x) = \int R'(x)\, dx$$

$$P(x) = R(x) - C(x)$$

Example 6.43

A business's marginal cost function is given by $C'(x) = 100x$ its marginal revenue function is $R'(x) = -0.02x^2 + 150x$. Find the profit function.

Solution: To find the profit function, we first integrate both functions:

$$C(x) = \int 100x\, dx = 50x^2 + c_1$$

$$R(x) = \int (-0.02x^2 + 150x)\, dx = -\frac{0.02}{3}x^3 + \frac{150}{2}x^2 + c_2$$

Now, subtract the total cost from the total revenue to obtain the profit function:

$$P(x) = R(x) - C(x)$$

$$P(x) = \left(-\frac{0.02}{3}x^3 + \frac{150}{2}x^2 + c_2\right) - \left(50x^2 + c_1\right)$$

$$P(x) = -\frac{0.02}{3}x^3 + \left(\frac{150}{2} - 50\right)x^2 + (c_2 - c_1)$$

The profit function is given by

$$P(x) = -\frac{0.02}{3}x^3 + 25x^2 + (c_2 - c_1),$$

where $(c_2 - c_1)$ represents the constant of integration for the profit function, as it is a constant we can write $(c_2 - c_1) = c$, therefore, the profit function is given by: $P(x) = -\frac{0.02}{3}x^3 + 25x^2 + c$. While it is often reasonable to assume that when zero item are produced or sold, that profit is zero, this need not be the case. For example, even if no items are produced or sold the profit could be less than zero since fixed costs still need to be paid.

Consumer Surplus and Producer Surplus

The definite integral has many applications in economics. In particular, it is essential in the definitions of consumer surplus and producer surplus.

Consumer surplus

We have discussed the price-demand function, also simply called the demand function, in both Chapters 1 and 5. This function is encountered in a fairly wide range of business related applications, particularly when finding the revenue function or in elasticity of demand questions, which are covered in Chapter 9. In essence, the price-demand function $p = D(x)$ relates the price p of some commodity with the quantity x of that commodity that consumers are willing to purchase for that price.

Suppose that a market price of p_0 has been established. This results in a corresponding quantity x_0 demanded at this price. In other words, $p_0 = D(x_0)$. In aggregate, consumers who value the product more than the market price p_0 experience a savings since they pay less than their maximum willingness to pay. This is not to say that any single consumer experiences this savings, but that all consumers together experience this savings.

When the demand function is $p = D(x)$, the **consumer surplus** is defined to be the area under the demand curve and above the price line $y = p_0$, from 0 to x_0. Mathematically, it is expressed as

$$\text{Consumer Surplus} = CS = \int_0^{x_0} D(x)\, dx - p_0 x_0.$$

Note that this formula could also have been written as the area between two curves. In some books the consumer surplus is defined to be

$$\text{Consumer Surplus} = CS = \int_0^{x_0} \left(D(x) - p_0 \right) dx.$$

Alternatively, when the demand function is given in the form $x = D(p)$, consumer surplus is calculated from the price p_0 up to the maximum price consumers are willing to pay, denoted by m_0, which is where the demand curve intersects the y-axis. The formula is

$$\text{Consumer Surplus} = CS = \int_{p_0}^{m_0} D(p)\, dp.$$

The consumer surplus is represented by the shaded region shown below.

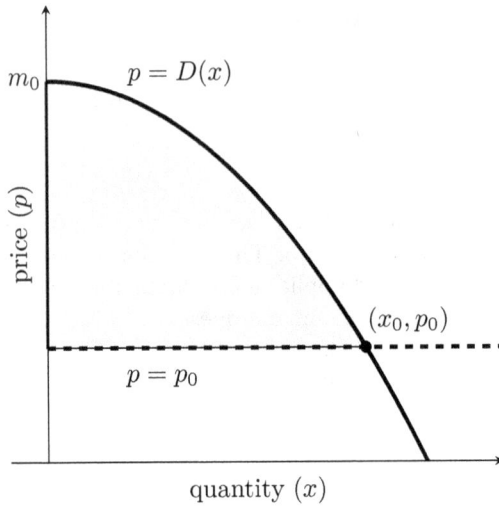

quantity (x)

Example 6.44

Consider a demand function given by $p = 30 - x - 0.5x^2$. Find the consumer surplus at $x_0 = 4$.

Solution: To calculate the consumer surplus at $x_0 = 4$, we integrate the demand curve from 0 to x_0 and then subtract the rectangle's area formed by x_0 and the price at x_0 denoted by $p_0 = D(x_0)$. First we find p_0 by substituting $x_0 = 4$ into the demand function,

$$p_0 = D(4) = 30 - 4 - 0.5(4)^2 = 30 - 4 - 8 = 18.$$

Then we find the consumer surplus when $x_0 = 4$ is

$$\text{CS} = \int_0^4 (30 - x - 0.5x^2)\, dx - 4 \cdot 18 = \frac{88}{3}.$$

Thus, the consumer surplus, when $x_0 = 4$, is approximately 29.33 dollars.

Example 6.45

If the demand function is $p = 24 + 2x - x^2$, find the consumer surplus at $x_0 = 5$.

Solution: To find the consumer surplus when $x_0 = 5$, we calculate the integral of the demand curve from 0 to x_0 and subtract the area of the rectangle formed by x_0 and the price at x_0, $p_0 = D(x_0)$. First, find p_0 by substituting $x_0 = 5$ into the demand function

$$p_0 = D(5) = 24 + 2(5) - (5)^2 = 9.$$

Then, the consumer surplus is:

$$\text{CS} = \int_0^5 (24 + 2x - x^2)\, dx - 5 \cdot 9 = \frac{175}{3}.$$

Therefore, the consumer surplus, when $x_0 = 5$, is approximately 58.33 dollars.

Example 6.46

Given the demand function for a luxury coffee $p = 50 - 2x$, calculate the consumer surplus when the market price p_0 is \$30.

Solution: First, determine x_0 by setting $p_0 = 30$ in the demand function,

$$30 = 50 - 2x_0 \implies x_0 = 10.$$

The consumer surplus is then calculated as

$$\text{CS} = \int_0^{10} (50 - 2x) - 30\, dx = 100$$

Thus, the consumer surplus when the market price is \$30 is \$100.

Example 6.47

Assume the demand function for online streaming subscriptions is $p = 40 - 0.5x$. Find the consumer surplus at a market price p_0 of \$20.

Solution: Find x_0 by substituting $p_0 = 20$ into the demand equation

$$20 = 40 - 0.5x_0 \implies x_0 = 40.$$

The consumer surplus is given by

$$\text{CS} = \int_0^{40} (40 - 0.5x) - 20\, dx = 400.$$

Therefore, the consumer surplus at a market price of \$20 is \$400.

Producer surplus

Supply functions are quite similar to demand functions, only from the perspective of the producers. Supply functions, usually denoted $p = S(x)$, represent the quantity of a product that producers are willing to sell at various prices.

Let p_0 be the market price and x_0 be the corresponding quantity supplied at this price. In aggregate, the producers who are willing to sell the product for less than the market price p_0 experience a gain since they receive more than their minimum willingness to sell. When the supply function is $p = S(x)$, producer surplus is calculated as the area above the supply curve and below the price line, from 0 to x_0. Mathematically, it is expressed as:

$$\text{Producer Surplus} = PS = p_0 x_0 - \int_0^{x_0} S(x)\, dx.$$

Again, this formula could have been written as the area between two curves. In some books the producer surplus is defined to be

$$\text{Producer Surplus} = PS = \int_0^{x_0} \Big(p_0 - S(x)\Big)\, dx.$$

Alternatively, when the supply function is given in the form $x = S(p)$, producer surplus is calculated from the price p_0 down to the minimum price producers are willing to accept, denoted by c_0, found where the supply curve intersects the y-axis. The formula is:

$$\text{Producer Surplus} = PS = \int_{c_0}^{p_0} S(p)\, dy.$$

The producer surplus is represented by the shaded region shown below.

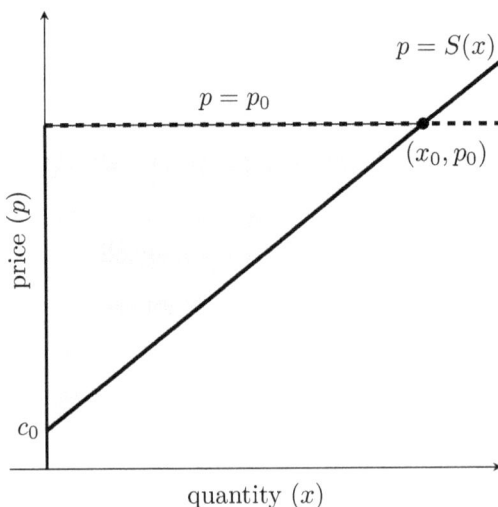

Example 6.48

If the supply function is $p = 2x$, find the producer surplus at $x_0 = 10$.

Solution: To find the producer surplus when $x_0 = 10$, calculate the integral of the supply curve from 0 to x_0 and subtract it from the area of the rectangle formed by x_0 and the price at $p_0 = S(x_0)$. First, find $s(x_0)$ by substituting $x_0 = 10$ into the supply function

$$p_0 = S(10) = 2(10) = 20.$$

Then, the producer surplus is

$$PS = 10 \cdot 20 - \int_0^{10} 2x \, dx = 200 - 100 = 100.$$

Therefore, The producer surplus, when $x_0 = 10$, is 100 dollars.

Example 6.49

Consider a supply function given by $p = 5 + 0.5x$. Find the producer surplus at $x_0 = 8$.

Solution: First, find p_0 by substituting $x_0 = 8$ into the supply function

$$p_0 = S(8) = 5 + 0.5(8) = 9.$$

The producer surplus when $x_0 = 8$ is

$$PS = 8 \cdot 9 - \int_0^8 (5 + 0.5x) \, dx = 16.$$

Thus, the producer surplus, when $x_0 = 8$, is approximately 16 dollars.

Example 6.50

Consider a market where the supply function of organic vegetables is given by $p = 2x + 4$, where x is the quantity in hundreds of kilograms, and p is the price per kilogram in dollars. If the market price is set at $20 per kilogram, calculate the producer surplus.

Solution: First, find x_0 by setting the market price $p_0 = 20$ in the supply function

$$20 = 2x + 4 \implies x = 8.$$

The producer surplus is the area above the supply curve and below the price line from 0 to x_0,

$$\text{PS} = \int_0^8 20 - (2x + 4)\, dx = 64.$$

Thus, the producer surplus is $64.

Example 6.51

Suppose the supply function for handcrafted pottery is given by $p = 3x^2$, where x is the number of pottery pieces, and p is the price per piece in dollars. Find the producer surplus when 5 pieces are supplied at a market price of $75 per piece.

Solution: First, verify that the supply function meets the market price for 5 pieces,

$$S(5) = 3(5)^2 = 75.$$

The producer surplus is the difference between the total revenue and the integral of the supply curve from 0 to x_0,

$$\text{PS} = \int_0^5 75 - 3x^2\, dx = 250.$$

Therefore, the producer surplus is $250.

6.6 PROBLEMS

Question 6.1 *Find the antiderivatives of the following functions:*

(a) $f(x) = x^3 + 5x^2 + 6x$

(b) $f(x) = -x^2 - x^3 - 3$

(c) $f(x) = x^4 - x^3 + 2$

(d) $f(x) = x^5(x - 1)$

(e) $f(x) = \frac{x^2 - 2x}{x}$

(f) $f(x) = 5^x + x^2 - e^x$

(g) $f(x) = \frac{a^2}{x}$

(h) $f(x) = \frac{2x}{1+x^2}$

(i) $f(x) = \frac{1}{3x+1}$

Question 6.2 *Find the definite integrals of the following functions:*

(a) $\int_0^1 x^3 + 5x^2 + 6x\, dx$

(b) $\int_{-1}^1 -x^2 - x^3 - 3\, dx$

(c) $\int_{-2}^2 x^4 - x^3 + 2\, dx$

(d) $\int_{-2}^2 x^5(x - 1)\, dx$

(e) $\int_2^3 \frac{x^2 - 2x}{x}\, dx$

(f) $\int_0^1 (5^x + x^2 - e^x)\, dx$

(g) $\int_{-2}^1 \frac{a^2}{x}$

(h) $\int_0^3 \frac{2x}{1+x^2}$

(i) $\int_0^3 \frac{1}{3x+1}$

Question 6.3 *Find the area enclosed by the* $f(x) = 4x - x^2$, *the x-axis*, $x = 0$, *and* $x = 2$.

Question 6.4 *Find the area enclosed by the* $f(x) = x^3 - x^2 - 5x$, *the x-axis*, $x = -1$, *and* $x = 1$.

Question 6.5 *Find the area enclosed by the* $f(x) = x^3$ *and* $g(x) = x$.

Question 6.6 *The weekly marginal revenue from the sale of x pairs of shoes is given by:* $R'(x) = 40 - 0.02x$, *where* $R(x)$ *is the revenue in dollars and* $R(x) = 0$ *when* $x = 0$.

1. *Find the revenue function.*

2. *Find the revenue from the sale of 1500 pairs of shoes.*

Question 6.7 *Find the revenue function* $R(q)$ *when the marginal revenue function is:* $R'(q) = 400 - 0.4q$ *and* $R(q) = 0$ *when* $q = 0$. *What is the revenue at a production level of 500 units?*

Question 6.8 *Given that the marginal cost for producing* q *units of a commodity is:* $C'(q) = 0.6q^2 + 2q$ *and the fixed costs are 2000.*

1. *Find the cost function* $C(q)$.

2. *Calculate the cost of producing 25 units.*

Question 6.9 *Given that the marginal cost for producing* x *thousand bottles of perfume per month is:* $C'(x) = 4x^2 - 25x + 50$.

1. *Find the cost function* $C(x)$ *if the fixed costs are $30,000.*

2. *Calculate the total cost of producing 4000 bottles of perfume per month.*

Question 6.10 *A manufacturing company has a marginal cost function for a raw material described by* $C'(y) = \frac{75000}{y} + y$, *where* y *is the quantity of raw material in kilograms. Determine the total cost function.*

Question 6.11 *A retail company purchases goods in bulk with the marginal cost function* $C'(x) = \frac{25000}{x} + 12x$. *Find the total cost function.*

Question 6.12 *A service provider has a marginal cost function for operations given by $C'(x) = \frac{4x^2 + 150}{x}$. Identify the total cost function.*

Question 6.13 *The marginal revenue from selling x units of a new technology product is modeled by $R'(x) = 300e^{-0.0002x}$. Determine the total revenue.*

Question 6.14 *A company sells smartphones at a price of $P(t) = 500 - 10t$ dollars per smartphone, and the quantity sold is given by $Q(t) = 100t - 5t^2$ smartphones over the time interval from $t = 0$ to $t = 20$ days. Calculate the total revenue during this period.*

Question 6.15 *A business's marginal cost function is given by $C'(x) = 80x$ dollars per unit, and its marginal revenue function is $R'(x) = -0.03x^2 + 120x$ dollars per unit. Find the profit function.*

Question 6.16 *If the demand function is $p = 8 - x$ and $x_0 = 4$, find the consumer surplus.*

Question 6.17 *The corresponding quantity and price, in perfect competition, are determined by the demand and supply functions $p = 10 - x^2$ and $p = 2 + x$ respectively. Determine the producer surplus.*

Question 6.18 *Assume the market for a product is in perfect competition with the supply function $y = 4 + 2x$ and the demand function $y = 16 - x$. If the market equilibrium quantity is $x_0 = 4$, determine the consumer surplus.*

Question 6.19 *Given the demand function $y = 12 - 2x$ and the market equilibrium quantity $x_0 = 3$, calculate the consumer surplus.*

Question 6.20 *Consider a market with the demand function $x = D(p) = 20 - 2p$ and the supply function $x = S(p) = 4p - 8$, where p represents the price in dollars and $D(p)$ and $S(p)$ represent the quantity demanded and supplied, respectively.*

1. *Determine the equilibrium price and quantity.*

2. *Calculate the consumer surplus at equilibrium.*

3. *Calculate the producer surplus at equilibrium.*

CHAPTER 7

More Integration Topics

We have presented a variety of integration rules that allow us to find the antiderivatives of a number of common functions. However, finding the antiderivatives of some functions requires more than just knowing an antiderivative rule. Some functions are more complicated and require us to use one of a number of integration techniques. In this chapter, we will introduce three of the most common integration techniques, and then see how they can be applied to business problems.

7.1 INTEGRATION BY SUBSTITUTION

We begin this chapter by introducing one of the most common integration techniques, called **integration by substitution**. On a fundamental level, integration by substitution relies on the chain rule. In a sense, we have already discussed this situation briefly in the last chapter when we provided the rule for indefinite integrals involving the chain rule. But we only provided rules for a couple of specific situations, namely those involving exponential functions with base e and the natural logarithm function. We want to take a deeper look that will allow us to address a wider range of situations. First, let us take a moment to review an example of the chain rule.

> **Example 7.1**
>
> Find the derivative of $f(x) = \frac{2}{3}\sqrt{(x+5)^3}$.
>
> Solution: We begin by finding the "inside" and "outside" functions. Hopefully, it is fairly clear that $u = x + 5$ is the "inside" function, while $y = \frac{2}{3}\sqrt{u^3} = \frac{2}{3}u^{\frac{3}{2}}$ is the "outside" function. The chain rule states that
>
> $$\frac{dy}{dx} = \frac{dy}{du}\frac{du}{dx} = \left(\frac{2}{3}u^{\frac{3}{2}}\right)' \cdot \left(x+5\right)' = u^{\frac{1}{2}} \cdot (1) = (x+5)^{\frac{1}{2}}.$$

DOI: 10.1201/9781003480235-7

Notice that the first derivative is taken with respect to the variable u and the second derivative is taken with respect to the variable x. In the final step, since $u = x + 5$, we substitute $x + 5$ in for the variable u. In summary, we have obtained $\left(\frac{2}{3}\sqrt{(x+5)^3}\right)' = \sqrt{x+5}$.

Next, we want to recognize a fairly straightforward fact, that the antiderivative of the derivative of a function is that function back, up to a constant of integration. We write that as

$$\int f'(x)\, dx = f(x) + c.$$

Thus, considering the last example, it is fairly clear that

$$\int \sqrt{x+5}\, dx = \frac{2}{3}\sqrt{(x+5)^3} + c.$$

Integration by substitution will allow us to find integrals like this directly. We will introduce this technique by continuing to use this example. Suppose we are asked to find the following integral:

$$\int \sqrt{x+5}\, dx.$$

We would begin by recognizing that we do know how to find the antiderivative of a function like $\sqrt{u} = u^{\frac{1}{2}}$ using the indefinite integral power rule. Thus, by substituting $u = x + 5$ we could rewrite the integral as follows:

$$\int \sqrt{x+5}\, dx = \int u^{\frac{1}{2}}\, dx.$$

But notice, our integrand is a function of u while dx indicates that the variable of integration is x, not u. How do we deal with this? Consider the substitution we made,

$$u = x + 5.$$

Taking the derivative of u with respect to x would give us

$$\frac{du}{dx} = u'(x) = (x+5)' = 1.$$

If we were to consider $\frac{du}{dx}$ as a fraction, we could multiply both sides by dx in order to clear the denominator, giving us

$$du = dx.$$

This gives us the relationship between the differential of u and the differential of x. Using this relationship, we can continue with our substitution,

$$\int \sqrt{x+5}\, dx = \int u^{\frac{1}{2}}\, dx = \int u^{\frac{1}{2}}\, du,$$

to obtain an integral that we know how to take.

$$\int u^{\frac{1}{2}} \, du = \frac{u^{\frac{1}{2}+1}}{\frac{1}{2}+1}$$

$$= \frac{u^{\frac{3}{2}}}{\frac{3}{2}}$$

$$= \frac{2}{3} u^{\frac{3}{2}}.$$

Our final step would be to then substitute back in the x using $u = x + 5$, to get

$$\int \sqrt{x+5} \, dx = \int u^{\frac{1}{2}} \, du = \frac{2}{3} u^{\frac{3}{2}} = \frac{2}{3} (x+5)^{\frac{3}{2}},$$

which is exactly what we expected. Notice, not only did we have to make a substitution to change the variable in the integrand, but we also needed to make a substitution to change the variable of integration. Let us look at a similar, though slightly more complicated, example.

Example 7.2

Find $\int x^2 \sqrt{x^3 + 7} \, dx$.

Solution: For this we let $u = u(x) = x^3 + 7$. We then notice that

$$\frac{du}{dx} = 3x^2 \implies du = 3x^2 \, dx \implies \frac{1}{3} \, du = x^2 \, dx.$$

With this we can do the following substitution and integration:

$$\int x^2 \sqrt{x^3 + 7} \, dx = \int \sqrt{u} \cdot x^2 \, dx$$

$$= \int \sqrt{u} \cdot \frac{1}{3} \, du$$

$$= \frac{1}{3} \int u^{\frac{1}{2}} \, du$$

$$= \frac{1}{3} \cdot \frac{2}{3} u^{\frac{3}{2}}$$

$$= \frac{2}{9} (x^3 + 7)^{\frac{3}{2}} + c.$$

Of course, the final answer needs to include a constant of integration, so we added the $+c$ to the final line.

Notice the overall strategy:

1. Let $u = u(x)$, where $u(x)$ is usually the "inside" part of the original composed function $f(u(x))$ in the integrand.
2. Find $du = u'(x)dx$.
3. Use substitution to convert the integral to one involving u only. In other words, the integral should be entirely in terms of u, not x.
4. Find the integral.
5. Replace u by $u(x)$ in the solution to obtain an expression in terms of x only.

Let us apply this strategy to another example.

Example 7.3

Find $\int (2x - 4)^3 \, dx$.

Solution: For this we let $u = u(x) = 2x - 4$. Using this we get

$$\frac{du}{dx} = u'(x) = 2 \implies du = 2 \, dx \implies \frac{1}{2} \, du = dx.$$

Now we make the substitutions and perform the integration.

$$\int (2x - 4)^3 \, dx = \int u^3 \, dx$$
$$= \int u^3 \cdot \frac{1}{2} \, du$$
$$= \frac{1}{2} \int u^3 \, du$$
$$= \frac{1}{2} \frac{u^4}{4}$$
$$= \frac{1}{8}(2x - 4)^4 + c.$$

With some practice it starts to become somewhat obvious which integrals are likely to be solved using the integration by substitution technique.

Example 7.4

Find $\int e^{-2x} \, dx$.

Solution: For this we let $u = u(x) = -2x$. Using this we get

$$\frac{du}{dx} = -2 \implies \frac{du}{-2} = dx.$$

We now make the substitution and perform the integration.

$$\int e^{-2x}\, dx = \int e^u\, \frac{du}{-2}$$

$$= -\frac{1}{2}\int e^u\, du$$

$$= -\frac{1}{2}e^u$$

$$= -\frac{1}{2}e^{-2x} + c.$$

Definite integrals are handled just as we would expect them to be.

Example 7.5

Find $\displaystyle\int_0^1 e^{-2x}\, dx$.

Solution: From the last example, we have $\int e^{-2x}\, dx = -\frac{1}{2}e^{-2x}$. When doing definite integrals we no longer need to worry about the constant of integration;

$$\int_0^1 e^{-2x}\, dx = \left[-\frac{1}{2}e^{-2x}\right]_0^1$$

$$= \left(-\frac{1}{2}e^{-2(1)}\right) - \left(-\frac{1}{2}e^{-2(0)}\right)$$

$$= -\frac{1}{2}e^{-2} + \frac{1}{2}e^0$$

$$= \frac{1 - e^{-2}}{2}.$$

Example 7.6

Find $\displaystyle\int \frac{2x + 4}{x^2 + 4x}\, dx$.

Solution: Let $u = u(x) = x^2 + 4x$. Using this we get

$$\frac{du}{dx} = 2x + 4 \implies du = (2x + 4)\, dx.$$

We now make the substitution and perform the integration.

$$\int \frac{2x+4}{x^2+4x}\,dx = \int \frac{(2x+4)}{u}\,dx$$

$$= \int \frac{1}{u}\,du$$

$$= \ln|u|$$

$$= \ln|x^2+4x| + c.$$

The last thing we would like to point out in this section is that not everything that looks like it could be integrated by substitution, can be. Sometimes this technique is insufficient. We will consider a very slightly tweaked version of the last example.

Example 7.7

Try to use integration by substitution for the integral $\int \frac{2x+6}{x^2+4x}\,dx$.

Solution: Again, the natural substitution to make would be $u = u(x) = x^2 + 4x$. And again, we would get

$$\frac{du}{dx} = 2x + 4 \implies du = (2x+4)dx.$$

But when we go to substitute back into the integral we discover that we are unable to

$$\int \frac{2x+6}{x^2+4x}\,dx = \int \frac{(2x+6)}{u}\,dx$$

$$= \int \frac{1}{u}\,(2x+6)\,dx,$$

which is not what we need in order to substitute du in. Thus, we can go no further.

7.2 INTEGRATION BY PARTS

In Chapter 3, we were given a rule for the integral of $\ln(x)$, namely that

$$\int \ln(x)\,dx = x\ln(x) - x.$$

For the moment we will disregard the constant c. In an example we then showed this rule was correct by taking the derivative of the right-hand side, which required the use of the product rule;

$$\left(x\ln(x) - x\right)' = \left(x\right)'\left(\ln(x)\right) + \left(x\right)\left(\ln(x)\right)' - 1$$

$$= \ln(x) + x \cdot \frac{1}{x} - 1$$
$$= \ln(x) + 1 - 1$$
$$= \ln(x).$$

It turns out that $\int \ln(x) dx$ is an example of an integral we can find using a method called **integration by parts**. We begin by taking a closer look at this particular example. Consider the function given by the product $x \ln(x)$. Clearly, taking the derivative of this function requires the use of the chain rule as follows:

$$\left(x \ln(x) \right)' = \left(x \right)' \left(\ln(x) \right) + \left(x \right) \left(\ln(x) \right)'$$
$$= 1 \cdot \ln(x) + x \cdot \frac{1}{x}$$
$$= \ln(x) + 1.$$

Suppose we then decided to integrate both sides of this equation. We would end up with

$$\int \left(x \ln(x) \right)' dx = \int \left(\ln(x) + 1 \right) dx$$
$$= \int \ln(x) \, dx + \int 1 \, dx.$$

The integral of the derivative of a function is simply that function back, up to a constant, which we are ignoring for the moment. In other words,

$$\int \left(x \ln(x) \right)' dx = x \ln(x).$$

It is also clear that

$$\int 1 \, dx = x.$$

Putting this together, we have that

$$\int \left(x \ln(x) \right)' dx = \int \ln(x) \, dx + \int 1 \, dx$$
$$\implies x \ln(x) = \int \ln(x) \, dx + x$$
$$\implies \int \ln(x) \, dx = x \ln(x) - x.$$

In essence, finding the integral of $\ln(x)$ required a clever use of the product rule.

To make this a little clearer, Let us turn our attention to the more general situation. Suppose we had the product of two functions, f and g. Using the product rule, we have

$$\left(fg\right)' = f'g + fg'.$$

Integrating both sides gives us

$$\int \left(fg\right)' \, dx = \int f'g \, dx + \int fg' \, dx,$$

which in turn gives us

$$fg = \int f'g \, dx + \int fg' \, dx.$$

This can be rewritten as

$$\int fg' \, dx = fg - \int f'g \, dx.$$

Integration by Parts:

Given functions f and g, we have

$$\int fg' \, dx = fg - \int f'g \, dx.$$

If limits of integration are included, we have

$$\int_a^b fg' \, dx = \left[fg\right]_a^b - \int_a^b f'g \, dx.$$

Integration by parts allows us to do a wider range of integration problems than we have encountered so far. Let us consider a few examples.

Example 7.8

Find $\int x^2 \ln(x) \, dx$.

Solution: By far the most challenging aspect of integration by parts is deciding what f and g' should be. It is helpful to remember that whatever you choose for f, you need to be able to find its derivative f' easily, and more importantly, whatever you choose for g', you need to be able to find its antiderivative easily. Here, let us choose

$$f = \ln(x) \quad \text{and} \quad g' = x^2$$

$$\implies f' = \frac{1}{x} \quad \text{and} \quad g = \frac{x^3}{3}.$$

Using the integration by parts formula we get

$$\int f g' \, dx = fg - \int f'g \, dx$$

$$\implies \int \ln(x) x^2 \, dx = \ln(x) \cdot \frac{x^3}{3} - \int \frac{1}{x} \cdot \frac{x^3}{3} \, dx$$

$$= \frac{x^3 \ln(x)}{x} - \frac{1}{3} \int x^2 \, dx$$

$$= \frac{x^3 \ln(x)}{x} - \frac{1}{3} \left(\frac{x^3}{3} \right)$$

$$= \frac{x^3 \ln(x)}{3} - \frac{x^3}{9}.$$

Of course, in the final answer we would need to add a constant of integration. One could easily check that this answer is correct by taking its derivative using the product rule.

A few hints to help us choose f and g' are in order. First, you need to be able to easily find the antiderivative of g'. It is also often helpful if g is somehow "simpler" (or at least not more "complex") than g' and if f' is "simpler" (or at least not more "complex") than f. Let us consider another example.

Example 7.9

Find $\displaystyle\int x e^x \, dx$.

Solution: We begin by trying to choose f and g'. If we let $f = x$, then it is fairly obvious that $f' = 1$ is somehow "simpler" than x. And if $g' = e^x$ then $g = e^x$ is certainly not more "complex" than g'. Thus we may guess

$$f = x \quad \text{and} \quad g' = e^x$$
$$\implies f' = 1 \quad \text{and} \quad g = e^x. \quad \text{("simpler")}$$

Applying the integration by parts formula we obtain

$$\int f g' \, dx = fg - \int f'g \, dx$$

$$\implies \int x e^x \, dx = x \cdot e^x - \int 1 \cdot e^x \, dx$$

$$= x \cdot e^x - \int e^x \, dx$$

$$= xe^x - e^x.$$

Again, the final answer would need to include the constant of integration c. Let us go ahead and check that our answer is correct, recalling that the derivative of the first term requires the product rule,

$$\left(xe^x - e^x\right)' = (x)'\left(e^x\right) + (x)\left(e^x\right)' - (e^x)'$$
$$= 1 \cdot e^x + x \cdot e^x - e^x$$
$$= xe^x.$$

Integration by parts problems can get quite tricky. For example, there are some problems that actually require us to apply integration by parts multiple times. Consider the following example:

Example 7.10

Find $\int x^2 e^x \, dx$.

Solution: We choose f and g',

$$f = x^2 \quad \text{and} \quad g' = e^x$$
$$\implies f' = 2x \quad \text{and} \quad g = e^x.$$

Applying integration by parts, we have

$$\int fg' \, dx = fg - \int f'g \, dx$$
$$\implies \int x^2 e^x \, dx = x^2 e^x - \int 2xe^x \, dx$$
$$= x^2 e^x - 2\left[\int xe^x \, dx\right].$$

We recognize $\int xe^x \, dx$ as the integral from the last example, that itself required integration by parts. Using the answer from the last example, we have

$$\int x^2 e^x \, dx = x^2 e^x - 2\left[\int xe^x \, dx\right]$$
$$= x^2 e^x - 2\left[xe^x - e^x\right]$$
$$= x^2 e^x - 2xe^x - 2e^x + c,$$

where we included the constant of integration in the last step. Thus, in order to do this integration we need to apply integration by parts twice.

Example 7.11

Find $\int \dfrac{xe^x}{(x+1)^2}\, dx$.

Solution: Since it is relatively straightforward to take the derivative of xe^x and find the antiderivative of $\frac{1}{(x+1)^2}$ using substitution, we choose

$$f = xe^x \quad \text{and} \quad g' = \frac{1}{(x+1)^2}$$

$$\implies f' = e^x + xe^x \quad \text{and} \quad g = \frac{-1}{x+1}.$$

This leads to

$$\int \frac{xe^x}{(x+1)^2}\, dx = xe^x \cdot \left(\frac{-1}{x+1}\right) - \int e^x(x+1) \cdot \left(\frac{-1}{x+1}\right) dx$$

$$= \frac{-xe^x}{x+1} + \int e^x\, dx$$

$$= \frac{-xe^x}{x+1} + e^x$$

$$= \frac{-xe^x}{x+1} + \frac{xe^x + e^x}{x+1}$$

$$= \frac{e^x}{x+1} + c,$$

where we added the constant of integration to the final answer.

7.3 INTEGRATION USING PARTIAL FRACTIONS

The method of partial fractions is a technique that allows us to integrate rational functions of the form

$$\frac{P(x)}{Q(x)}$$

where $P(x)$ and $Q(x)$ are polynomials. We will begin by working through a fairly simple example that will illustrate the general idea. Once we have done that we will give some more details about the overall technique and then work on a variety of other examples.

Suppose we wanted to find the following integral:

$$\int \frac{1}{x^2 + 8x + 15}\, dx.$$

What we want to do is transform the integrand into something that we already know how to handle. For this we will use the process of **partial fractions**.

In essence, we want to rewrite the integrand as a sum of fractions that we know how to integrate already. In order to do this we need to first factor the denominator,

$$x^2 + 8x + 15 = (x+3)(x+5),$$

thereby allowing us to rewrite the integrand as

$$\frac{1}{x^2 + 8x + 15} = \frac{1}{(x+3)(x+5)}.$$

Next, we want to find values A and B such that

$$\frac{1}{(x+3)(x+5)} = \frac{A}{(x+3)} + \frac{B}{(x+5)}.$$

This is not difficult to do, but it does require some algebra. We begin by clearing the denominator to get

$$1 = A(x+5) + B(x+3)$$
$$\implies \quad 1 = Ax + 5A + Bx + 3B$$
$$\implies \quad 0x + 1 = (A+B)x + (5A+3B)$$
$$\implies \quad 0 = A + B$$
$$1 = 5A + 3B,$$

which is a system of two equations with two unknowns. Notice how we equated the coefficient of the variable x on the left, which was zero, with the coefficient of the variable x on the right, which was $A + B$. Similarly, we equated the constant 1 on the left with the constant $5A + 3B$ on the right. Solving this system is quite straight-forward; the first equation gives us $B = -A$. Plugging $-A$ in for B in the second equation gives us $1 = 2A$, or $A = \frac{1}{2}$. Thus $B = -\frac{1}{2}$. We can then rewrite the integrand as

$$\frac{1}{x^2 + 8x + 15} = \frac{1}{(x+3)(x+5)} = \frac{1/2}{(x+3)} + \frac{-1/2}{(x+5)}.$$

This gives us

$$\int \frac{1}{x^2 + 8x + 15}\, dx = \int \frac{1/2}{(x+3)} + \frac{-1/2}{(x+5)}\, dx$$
$$= \frac{1}{2}\int \frac{1}{(x+3)}\, dx - \frac{1}{2}\int \frac{1}{(x+5)}\, dx.$$

In order to do the two integration problems you would need to use substitution. For the first integral we let $u = x + 3$, which we use to find $du = dx$, so we have

$$\int \frac{1}{(x+3)}\, dx = \int \frac{1}{u}\, du = \ln|u| = \ln|x+3|.$$

The second integral is similar. Thus, as a final answer, we have

$$\int \frac{1}{x^2 + 8x + 15} \, dx = \frac{1}{2} \ln |x + 3| - \frac{1}{2} \ln |x + 5| + c,$$

where we added the constant of integration in the last step.

The general idea for integration by parts is that we can decompose integrands that are rational functions $\frac{P(x)}{Q(x)}$ into a sum of simpler rational functions whose antiderivatives we can more readily find. We now provide the details on how this decomposition works in general.

1. Given $\frac{P(x)}{Q(x)}$, if the degree of polynomial $P(x)$ is greater than or equal to the degree of polynomial $Q(x)$, then first perform long division. The quotient polynomial can be integrated directly and the method of partial fraction can be used on the remainder.

2. If the denominator $Q(x)$ is the product of linear terms of the form $x - c$, then the method of partial fractions uses terms of form

$$\frac{A}{x - c}.$$

3. If the denominator $Q(x)$ contains any repeated linear factors, $(x - c)^n$, then use terms of the form

$$\frac{A_1}{(x - c)} + \frac{A_2}{(x - c)^2} + \cdots + \frac{A_n}{(x - c)^n}.$$

4. If the denominator $Q(x)$ contains any unfactorable quadratics $q(x)$, then use use terms of the form

$$\frac{Ax + B}{q(x)}.$$

We have already looked at an example where the denominator $Q(x)$ is the product of two linear factors. We will consider one more example of this form.

Example 7.12

Find $\int \dfrac{2x + 1}{x^2 - 1} \, dx$.

Solution. We start out by factoring the denominator of the integrand $x^2 - 1 = (x + 1)(x - 1)$. We then rewrite the integrand as a sum of partial fractions,

$$\frac{2x + 1}{(x + 1)(x - 1)} = \frac{A}{(x + 1)} + \frac{B}{(x - 1)}.$$

Clearing the denominator and doing some algebra gives us

$$2x + 1 = A(x - 1) + B(x + 1)$$
$$\implies \quad 2x + 1 = Ax - A + Bx + B$$
$$\implies \quad 2x + 1 = (A + B)x + (-A + B)$$
$$\implies \quad 2 = A + B$$
$$1 = -A + B,$$

which again is a system of two equations with two unknowns. Solving the first equation for B and substituting that into the second equation gives $A = \frac{1}{2}$. Using this we find $B = \frac{3}{2}$. Thus, we have

$$\int \frac{2x + 1}{x^2 - 1}\, dx = \int \frac{1/2}{(x + 1)} + \frac{3/2}{(x - 1)}\, dx$$
$$= \frac{1}{2} \int \frac{1}{(x + 1)}\, dx + \frac{3}{2} \int \frac{1}{(x - 1)}\, dx$$
$$= \frac{1}{2} \ln|x + 1| + \frac{3}{2} \ln|x - 1| + c,$$

where the constant of integration was included in the last step.

We will now consider an example where $Q(x)$ contains repeated linear factors.

Example 7.13

Find $\int \dfrac{2x^2 - x + 4}{(x - 2)^2(x + 3)}\, dx.$

Solution: Notice, in the denominator of the integrand the factor $(x - 2)$ is repeated twice. We also have the factor $(x + 3)$. If we were to multiply out the denominator, we would see that it has degree three, while the numerator only has degree two, thus we do not need to worry about step (1). The repeated term $(x - 2)^2$ requires a term of the form

$$\frac{A}{(x - 2)} + \frac{B}{(x - 2)^2},$$

while the linear factor $(x + 3)$ requires the term

$$\frac{C}{(x + 3)}.$$

We combine these terms to get

$$\frac{2x^2 - x + 4}{(x-2)^2(x+3)} = \frac{A}{(x-2)} + \frac{B}{(x-2)^2} + \frac{C}{(x+3)}.$$

Next, we clear the denominator, multiply out, and combine like terms;

$$\begin{aligned}
2x^2 - x + 4 &= A(x-2)(x+3) + B(x+3) + C(x-2)^2 \\
&= A(x^2 + x - 6) + B(x+3) + C(x^2 - 4x + 4) \\
&= Ax^2 + Ax - 6A + Bx + 3B + Cx^2 - 4Cx + 4C \\
&= (A+C)x^2 + (A+B-4C)x + (-6A+3B+4C).
\end{aligned}$$

This gives us the following system of three equations in three unknowns:

$$\begin{aligned}
2 &= A + C \\
-1 &= A + B - 4C \\
4 &= -6A + 3B + 4C.
\end{aligned}$$

It is cumbersome, though straight-forward, to find $A = 1$, $B = 2$, and $C = 1$, yielding

$$\int \frac{2x^2 - x + 4}{(x-2)^2(x+3)}\, dx$$

$$= \int \frac{1}{(x-2)}\, dx + 2\int \frac{1}{(x-2)^2}\, dx + \int \frac{1}{(x+3)}\, dx.$$

All three of these integrals can be solved using substitution. We have already done integrals similar to the first and third, so will focus on the second. Letting $u = x - 2$, we have $du = dx$, which gives us

$$\int \frac{1}{(x-2)^2}\, dx = \int u^{-2}\, du$$

$$= \frac{u^{-2+1}}{-2+1} = \frac{u^{-1}}{-1}$$

$$= -\frac{1}{u} = -\frac{1}{x-2}.$$

Putting this all together, we have

$$\int \frac{2x^2 - x + 4}{(x-2)^2(x+3)}\, dx = \ln|x-2| - \frac{2}{x-2} + \ln|x+3| + c.$$

Example 7.14

Find $\int \dfrac{x^2 + 2x - 2}{(x^2 + 2)(x + 1)}\, dx$.

Solution: We begin by noting that the term $x^2 + 2$ is an unfactorable quadratic, at least if we are only considering real numbers and not complex numbers. Thus, the factor $x^2 + 2$ requires a term of the form

$$\frac{Ax + B}{x^2 + 2}$$

and the factor $x + 1$ requires a term of the form

$$\frac{C}{x + 1}.$$

We combine these terms to get

$$\frac{x^2 + 2x - 2}{(x^2 + 2)(x + 1)} = \frac{Ax + B}{x^2 + 2} + \frac{C}{x + 1}.$$

Clearing the denominator gives us

$$x^2 + 2x - 2 = (Ax + B)(x + 1) + C(x^2 + 2)$$
$$= Ax^2 + Ax + Bx + B + Cx^2 + 2C$$
$$= (A + C)x^2 + (A + B)x + (B + 2C).$$

This gives us the following system of three equations in three unknowns:

$$1 = A + C$$
$$2 = A + B$$
$$-2 = B + 2C.$$

Again, solving for these unknowns is cumbersome but not difficult. We obtain $A = 2$, $B = 0$, and $C = -1$. Thus, we have

$$\int \frac{x^2 + 2x - 2}{(x^2 + 2)(x + 1)}\, dx = \int \frac{2x}{x^2 + 2}\, dx - \int \frac{1}{x + 1}\, dx.$$

Each of theses can be integrated using substitution giving us a final answer of

$$\int \frac{x^2 + 2x - 2}{(x^2 + 2)(x + 1)}\, dx = \ln|x^2 + 2| - \ln|x + 1| + c.$$

7.4 APPLICATIONS OF INTEGRATION IN BUSINESS AND ECONOMICS

It is of course obvious that the integration techniques we have just learned can be applicable in a variety of business and economics applications. We begin with some examples that require substitution.

Example 7.15

The sales department estimates that the monthly sales of a new product during its first three years will be given by $1000 - 750e^{-0.1t}$, where $0 \leq t \leq 36$. Find an expression for total sales at time t.

Solution: If we let $S(t)$ represent the total sales for $0 \leq t \leq 36$, then we have

$$S'(t) = 1000 - 750e^{-0.1t}$$

so

$$S(t) = \int 1000 - 750e^{-0.1t} \, dt$$

$$= \int 1000 \, dt - \int 750e^{-0.1t} \, dt.$$

The first integral is easy, but the second integral requires us to use the method of substitution. Letting $u = -0.1t$, we have $du = -0.1dt$ giving us

$$S(t) = 1000t - 750 \int e^u \frac{du}{-0.1}$$

$$= 1000t - \frac{750}{-0.1} \int e^u \, du$$

$$= 1000t + 7500e^u$$

$$= 1000t + 7500e^{-0.1t} + c.$$

We included the constant in the last line. It is reasonable to expect sales to be zero at time zero, so $S(0) = 0$, allowing us to solve for the constant $c = -7500$. Thus the expression for total sales is given by

$$S(t) = 1000t + 7500e^{-0.1t} - 7500.$$

Example 7.16

A marketing analysis predicts that the growth in the number of subscribers for a new online service can be modeled by the function:

$$G(t) = 1500 \left(1 - e^{-0.3t}\right)$$

where $G(t)$ represents the number of new subscribers after t months. Calculate the total number of subscribers gained over the first two years.

Solution: If $N(t)$ is the total number of subscribers from $0 \le t \le 24$ months then $N'(t) = G(t)$, which gives us

$$N'(t) = 1500 \left(1 - e^{-0.3t}\right).$$

To find $N(t)$:

$$N(t) = \int 1500 \left(1 - e^{-0.3t}\right) \, dt$$

$$= \int 1500 \, dt - \int 1500 e^{-0.3t} \, dt.$$

The integral for the first term is straightforward. The second term requires integration by substitution. Substituting for the exponential term, let

$$u = -0.3t \quad \Longrightarrow \quad du = -0.3dt \quad \Longrightarrow \quad dt = \frac{du}{-0.3},$$

giving us

$$N(t) = 1500t - 1500 \int e^u \frac{du}{-0.3}$$

$$= 1500t - \frac{1500}{-0.3} \int e^u \, du$$

$$= 1500t + 5000 e^u$$

$$= 1500t + 5000 e^{-0.3t} + c.$$

Assuming $N(0) = 0$ (i.e., no subscribers at the start), solve for c:

$$N(0) = 1500 \times 0 + 5000 e^0 + c \implies c = -5000.$$

Thus, the total number of subscribers over two years (24 months) is given by:

$$N(24) = 1500(24) + 5000 e^{-0.3(24)} - 5000 \approx 31{,}004.$$

Example 7.17

A natural resource is being depleted following a power law model:

$$R(t) = \frac{500}{(1+t)^2}$$

where $R(t)$ is the rate of depletion in units per year, and t is time in years. The task is to find the total depletion over the first 10 years.

Solution: Let $D(t)$ represent the total depletion for $0 \le t \le 10$,

$$D'(t) = \frac{500}{(1+t)^2}.$$

We aim to integrate this rate of depletion to find the total depletion, $D(t)$, over the given time period. Using substitution, let $u = 1 + t$ and $du = dt$. Changing the limits of integration based on the substitution:

$$\text{When } t = 0, \quad u = 1,$$
$$\text{When } t = 10, \quad u = 11.$$

The integral becomes:

$$D(t) = \int \frac{500}{u^2} \, du.$$

Solving the integral within the new limits:

$$D(t) = \int_0^{10} \frac{500}{(1+t)^2} \, dt$$

$$= \int_1^{11} \frac{500}{u^2} \, du$$

$$= 500 \left[-\frac{1}{u} \right]_1^{11}$$

$$= 500 \left(-\frac{1}{11} + \frac{1}{1} \right)$$

$$= \frac{5000}{11}.$$

Therefore, the total resource depleted over the first 10 years is $\frac{5000}{11}$ units.

The next example is interesting in that it first requires substitution, but then we need to use integration by partial fractions.

Example 7.18

The marketing team predicts that the adoption of a new technology follows a logistic growth model, represented by the function:

$$R(t) = \frac{2000}{1 + 25e^{-0.5t}}$$

where $R(t)$ is the rate of adoption per month, and t is time in months. Find an equation for the total number of adopters over the first 24 months.

Solution: Let $A(t)$ represent the total adopters from $t = 0$ to $t = 24$. Thus $A'(t) = R(t)$, so

$$A'(t) = \frac{2000}{1 + 25e^{-0.5t}}.$$

To integrate we will use substitution. Let

$$u = 1 + 25e^{-0.5t}$$

$$\implies \frac{du}{dt} = -12.5e^{-0.5t}$$

$$\implies dt = \frac{du}{-12.5e^{-0.5t}}.$$

To continue further we need to use $e^{-0.5t} = \frac{u-1}{25}$, which gives us

$$\implies dt = \frac{du}{-12.5\left(\frac{u-1}{25}\right)}$$

$$\implies dt = \frac{-25\,du}{12.5(u-1)}$$

$$\implies dt = \frac{-2\,du}{u-1}.$$

Substituting into our equation, we have

$$A(t) = \int \frac{2000}{1 + 25e^{-0.5t}}\,dt$$

$$= \int \frac{2000}{u}\,dt$$

$$= \int \frac{2000}{u} \cdot \frac{-2\,du}{u-1}$$

$$= -4000 \int \frac{1}{u(u-1)}\,du.$$

At this point we need to use integration by partial fractions.

$$\frac{1}{u(u-1)} = \frac{A}{u} + \frac{B}{u-1}$$

$$\Longrightarrow \frac{1}{u(u-1)} = \frac{A(u-1)}{u(u-1)} + \frac{Bu}{u(u-1)}$$

$$\Longrightarrow 1 = Au - A + Bu$$

$$\Longrightarrow 0u + 1 = (A+B)u - A$$

$$\Longrightarrow A = -1 \quad \text{and} \quad B = 1.$$

Thus,

$$A(t) = -4000 \int \frac{-1}{u} + \frac{1}{u-1}\, du$$

$$= -4000 \left[-\ln|u| + \ln|u-1| \right] + c.$$

We now need to substitute back in u;

$$A(t) = 4000 \ln \left|1 + 25e^{-0.5t}\right| - 4000 \ln \left|25e^{-0.5t}\right| + c.$$

Since $A(0) = 0$ we can solve for c to get $c = -4000 \ln\left(\frac{26}{25}\right)$, giving us

$$A(t) = 4000 \ln\left(1 + 25e^{-0.5t}\right) - 4000 \ln\left(25e^{-0.5t}\right) - 4000 \ln\left(\frac{26}{25}\right).$$

The number of adopters over the first 24 months is $A(24) \approx 34{,}968$.

Example 7.19

A company applies a depreciation formula over the economic lifespan of an asset. Total depreciation at time T is given by

$$D(T) = \int_0^T \frac{10000}{t^2 + 10t + 25}\, dt$$

where t is time in years. Calculate the formula for the total depreciation of this asset at time T.

Solution: First factor the denominator, $t^2 + 10t + 25 = (t+5)^2$. This might look like we need integration by partial fractions, but substitution will actually work. Letting $u = t + 5$ we get $du = dt$, thus

$$\int \frac{10000}{(t+5)^2}\, dt = 10000 \int u^{-2}\, du = 10000 \cdot (-u^{-1}) = -\frac{10000}{t+5}.$$

Giving us

$$D(T) = \left[-\frac{10000}{t+5}\right]_0^T = 2000 - \frac{10000}{T+5}.$$

The next few examples require integration by parts.

Example 7.20

Consider a machine whose rate of depreciation over time following a function related to its usage, $r(t) = te^{0.1t}$, where r is the depreciation rate in thousands of dollars and t is time in years. Calculate the total depreciation over the first 3 years.

Solution: To find the total depreciation from $t = 0$ to $t = 3$, we integrate the depreciation rate $r(t)$ over this interval:

$$\text{Total Depreciation} = \int_0^3 te^{0.1t}\, dt.$$

We can solve this integral using integration by parts, by letting

$$f = t \quad \text{and} \quad g' = e^{0.1t}$$
$$\implies f' = 1 \quad \text{and} \quad g = \frac{e^{0.1t}}{0.1}.$$

Using the formula,

$$\int fg'\, dx = fg - \int f'g\, dx$$
$$\implies \int te^{0.1t}\, dt = t \cdot \frac{e^{0.1t}}{0.1} - \int 1 \cdot \frac{e^{0.1t}}{0.1}\, dt$$
$$= 10te^{0.1t} - 10 \int e^{0.1t}\, dt$$
$$= 10te^{0.1t} - 100e^{0.1t}.$$

Taking the definite integral, we have

$$\int_0^3 te^{0.1t}\, dt = \left[10te^{0.1t} - 100e^{0.1t}\right]_0^3$$
$$= \left(10 \cdot 3 \cdot e^{0.3} - 100 \cdot e^{0.3}\right) - \left(10 \cdot 0 \cdot e^0 - 100 \cdot e^0\right)$$
$$= -70e^{0.3} + 100 \approx 5.50988.$$

The total depreciation over the first three years is $5,509.88.

Example 7.21

A marketing department estimates that the annual rate of production for a new product, in thousands of units, will be given by $r(t) = 50te^{-0.2t}$, where t is in years since product rollout. (Note that the production rate in year 0 is 0.) Find an expression that describes total product production in year t.

Solution: Since the derivative of total product production, $T'(t)$ is the rate of the product production, $r(t)$, in order to get the total product production function we need to find the antiderivative of the rate of production. In other words, we are interested in finding

$$T(t) = \int 50te^{-0.2t}\, dt.$$

We can solve this integral using integration by parts, by letting

$$f = 50t \quad \text{and} \quad g' = e^{-0.2t}$$

$$\Longrightarrow f' = 50 \quad \text{and} \quad g = \frac{-e^{-0.2t}}{0.2}.$$

Thus, we have

$$\int fg'\, dx = fg - \int f'g\, dx$$

$$\Longrightarrow \int 50te^{-0.2t}\, dt = 50t \cdot \frac{-e^{-0.2t}}{0.2} - \int 50 \cdot \frac{-e^{-0.2t}}{0.2}\, dx$$

$$= -250te^{-0.2t} + 250 \int e^{-0.2t}\, dx$$

$$= -250te^{-0.2t} + 250 \cdot \frac{-e^{-0.2t}}{0.2}$$

$$= -250te^{-0.2t} - 1250e^{-0.2t} + c.$$

Knowing that $T(0) = 0$ we easily find that $c = 1250$. Thus, total production is given by

$$T(t) = 1250 - 250te^{-0.2t} - 1250e^{-0.2t}.$$

Example 7.22

A company purchases a new piece of equipment for \$120,000 and estimates that the value of the equipment decreases at a rate given by the function $r(t) = 10{,}000te^{-0.1t}$ in dollars per year, where t is the time in years. Find an expression for the total value of the equipment after t years.

Solution: Since the derivative of the total value of the equipment, $V'(t)$, is the rate of depreciation, $r(t)$, we seek to find the total value by integrating the depreciation rate. Thus,

$$V(t) = 120{,}000 - \int 10{,}000te^{-0.1t}\, dt$$

$$= 120{,}000 - 10{,}000 \int te^{-0.1t}\, dt.$$

Applying integration by parts, let:

$$f = t \quad \text{and} \quad g' = e^{-0.1t}$$
$$f' = 1 \quad \text{and} \quad g = -10e^{-0.1t}.$$

Then,

$$\int fg'\, dx = fg - \int f'g\, dx$$

$$\Longrightarrow \int t \cdot e^{-0.1t}\, dt = -10te^{-0.1t} - \int -10e^{-0.1t}\, dt$$

$$= -10te^{-0.1t} + 10 \int e^{-0.1t}\, dt$$

$$= -10te^{-0.1t} + 10 \cdot -10e^{-0.1t}$$

$$= -10te^{-0.1t} - 100e^{-0.1t} + c.$$

Combining with what we had before we find

$$V(t) = 120{,}000 - 10{,}000 \left[-10te^{-0.1t} - 100e^{-0.1t} + c \right]$$
$$= 120{,}000 + 100{,}000te^{-0.1t} + 1{,}000{,}000e^{-0.1t} + c,$$

where we simply relabel $10{,}000c$ as c. Since we know the value of the machine at time $t = 0$ is $120{,}000$ we can solve for c;

$$V(0) = 120{,}000 + 100{,}000te^{-0.1t} + 1{,}000{,}000e^{-0.1t} + c = 120{,}000$$
$$\Longrightarrow 120{,}000 + 0 + 1{,}000{,}000 + c = 120{,}000$$

giving us $c = -1{,}000{,}000$. Thus, we have

$$V(t) = 120{,}000 + 100{,}000te^{-0.1t} + 1{,}000{,}000e^{-0.1t} - 1{,}000{,}000$$
$$= 100{,}000te^{-0.1t} + 1{,}000{,}000e^{-0.1t} - 880{,}000.$$

Example 7.23

A company plans to increase the sales price of its product linearly from $20 to $50 over 10 years. The rate of sales is expected to decrease exponentially with the price increase, modeled by $s(t) = 1000e^{-0.05t}$, where $s(t)$ is the rate of sales in units per year and t is time in years. Find the total revenue function.

Solution: Since price increases linearly from $20 to $50 over a ten year period, the price is given by $p(t) = 20 + 3t$ for $0 \leq t \leq 10$. The revenue function $R(t)$ can be found by integrating the product of price and sales rate;

$$R(t) = \int (20 + 3t) \cdot 1000e^{-0.05t} \, dt.$$

Letting

$$f = 20 + 3t \quad \text{and} \quad g' = 1000e^{-0.05t}$$
$$f' = 3 \quad \text{and} \quad g = -20{,}000e^{-0.05t}.$$

we then use integration by parts, $\int f g' \, dx = fg - \int f'g \, dx$, which gives us

$$\int (20 + 3t) \cdot 1000e^{-0.05t} \, dt$$

$$= (20 + 3t) \cdot \left(-20{,}000e^{-0.05t}\right) - \int 3 \cdot -20{,}000 \cdot e^{-0.05t} \, dt$$

$$= -20{,}000(20 + 3t)e^{-0.05t} + 60{,}000 \int e^{-0.05t} \, dt$$

$$= -20{,}000(20 + 3t)e^{-0.05t} - 1{,}200{,}000 \cdot e^{-0.05t} + c.$$

Without being given some additional information we are unable to solve for the constant c, thus, this is as far as we can go.

This next example is particularly interesting because it requires integration by parts, by partial fractions, and by substitution.

Example 7.24

Suppose a company's marketing department estimates that the rate of the effect of a new advertising campaign on product sales, measured in thousands of units per month, follows a logarithmic decay represented by

$$r(t) = 100 \ln(1 + 0.5t)$$

where t is the time in months since the beginning of the campaign. Calculate the total impact on sales over the first 12 months.

Solution: Since the derivative of total impact, $I'(t)$, is the rate of sales impact $r(t)$, we need to find the antiderivative:

$$I(t) = \int 100 \ln(1 + 0.5t) \, dt.$$

To solve this, we use integration by parts. Let

$$f = \ln(1 + 0.5t) \quad \text{and} \quad g' = 100$$

$$f' = \frac{0.5}{1 + 0.5t} \quad \text{and} \quad g = 100t.$$

Applying integration by parts, $\int fg' \, dx = fg - \int f'g \, dx$, we get

$$\int 100 \ln(1 + 0.5t) \, dt = 100t \ln(1 + 0.5t) - \int \frac{50t}{1 + 0.5t} \, dt.$$

The integral on the right-hand side will require integration by partial fractions. But notice, in this case, the numerator has the same degree as the denominator, which means we need to perform long polynomial division. Instead of doing that, which we have not covered in this book, we will instead simply recognize that the integrand, after division, would have the form

$$\frac{50t}{1 + 0.5t} = A + \frac{B}{1 + 0.5t}.$$

We need to solve for A and B;

$$\frac{50t}{1 + 0.5t} = A + \frac{B}{1 + 0.5t}$$

$$\implies \frac{50t}{1 + 0.5t} = \frac{A(1 + 0.5t)}{1 + 0.5t} + \frac{B}{1 + 0.5t}$$

$$\implies 50t = A + 0.5At + B$$

$$\implies 50t + 0 = 0.5At + (A + B)$$

$$\implies A = 100 \quad \text{and} \quad B = -100.$$

Which allows us to perform the necessary integration,

$$\int \frac{50t}{1 + 0.5t} \, dt = \int 100 - \frac{100}{1 + 0.5t} \, dt$$

$$= 100t - \frac{100}{0.5} \ln(1 + 0.5t),$$

where, technically, we would need a substitution for the second integral on the right. Putting all of this back into the first integral, we obtain

$$I(t) = 100t \ln(1 + 0.5t) - 100t + 200 \ln(1 + 0.5t) + c.$$

Calculating $I(12) - I(0) = 1524.27$ yields the total impact on sales over the first 12 months.

Example 7.25

A business takes out a decreasing term loan with a repayment interest rate formula modeled by:

$$r(t) = \frac{3000}{t^2 + 13t + 40}$$

where x represents the time in months since the loan was issued. Find the total interest accumulated by time T.

Solution: Accumulated interest at time T is given by

$$I(T) = \int_0^T \frac{3000}{t^2 + 13t + 40} \, dt.$$

This will require integration by partial fractions. To find the integral, we first, factor the denominator as

$$t^2 + 13t + 40 = (t + 5)(t + 8).$$

Then we decompose the integrand into partial fractions:

$$\frac{3000}{(t + 5)(t + 8)} = \frac{A}{t + 5} + \frac{B}{t + 8}$$
$$\implies 3000 = A(t + 8) + B(t + 5)$$
$$\implies A = 1000 \quad \text{and} \quad B = -1000.$$

Thus we get

$$\int \left(\frac{1000}{t + 5} - \frac{1000}{t + 8} \right) dx = 1000 \Big(\ln|t + 5| - \ln|t + 8| \Big).$$

Using the Fundamental Theorem of Calculus, we have

$$I(T) = 1000 \Big(\ln|T + 5| - \ln|T + 8| - \ln|5| + \ln|8| \Big)$$
$$= 1000 \ln \left(\frac{8(T + 5)}{5(T + 8)} \right).$$

Integrals are also used in other applications where we compute probabilities. While this is not a book in probability, this is one area where integration is frequently used in economics theory, thus it is useful to see a few examples of this nature. Let us assume that individual incomes in a given population are greater than or equal to A (where A denotes the minimum income). Let $F(x)$ denote the proportion of individuals with an income less than x. If F is differentiable, with derivative f, then we have

$$F(x) = \int_A^x f(t)\, dt.$$

For $A < a < b$, the proportion of individuals with an income between a and b is

$$\int_a^b f(t)\, dt.$$

If we select an individual at random from this population, the probability that his/her income X is between a and b is

$$P(a \leq X \leq b) = \int_a^b f(t)\, dt = F(b) - F(a).$$

This integral quantifies the total area under the function f curve from a to b and represents the probability that income X takes a value within that interval.

We can also consider the probability that the income X is higher than a specific value c, which is given by:

$$P(X \geq c) = \int_c^\infty f(t)\, dt = 1 - \int_0^c f(t)\, dt = 1 - F(c).$$

Furthermore, the expected value of income denoted $E(X)$, which provides the average income, is given by the integral:

$$E(X) = \int_0^\infty t f(t)\, dt.$$

Example 7.26

Suppose that the income distribution function $f(x)$ for a small city is given by

$$f(x) = \begin{cases} 0 & \text{if } x < 500 \\ \frac{1}{44.72\sqrt{x}} & \text{if } x \geq 500 \end{cases}$$

What is the probability that a randomly chosen individual earns between 1000 and 2000?

Solution: The probability that a randomly chosen individual earns between 1000 and 2000 is calculated using the integral of the income function over this interval:

$$P(1000 \le X \le 2000) = \int_{1000}^{2000} \frac{1}{44.72\sqrt{x}}\, dx.$$

Using substitution, let $u = \sqrt{x}$, then $du = \frac{1}{2\sqrt{x}}\, dx$ or $dx = 2u\, du$.

$$\int \frac{1}{44.72} \cdot \frac{1}{u} \cdot 2u\, du = \frac{2}{44.72} \int 1\, du = \frac{2}{44.72} u = \frac{2}{44.72}\sqrt{x}.$$

Now, doing the definite integral, we have

$$\int_{1000}^{2000} \frac{1}{44.72\sqrt{x}}\, dx = \frac{2}{44.72}\left(\sqrt{2000} - \sqrt{1000}\right) \approx 0.586.$$

Example 7.27

Let us assume that the income function $f(x)$ is given by

$$f(x) = \begin{cases} 0 & \text{if } x < 500 \\ \frac{1}{155\sqrt{x}} & \text{if } 500 \le x \le 10000 \end{cases}$$

Here, the minimum income is $500 and the maximum income is $ 10,000.

1. What is the probability that a randomly chosen individual earns
   between $1000 and $2000?
2. What is the probability that a randomly chosen individual earns
   more than $3000?

Solution:

1. The probability that a randomly chosen individual earns between
   $1000 and $2000 is calculated by the integral of the income function
   over this range:

$$P(1000 \le X \le 2000) = \int_{1000}^{2000} \frac{1}{155\sqrt{x}}\, dx.$$

Using substitution, let $u = \sqrt{x}$, then $du = \frac{1}{2\sqrt{x}}\, dx$ or $dx = 2u\, du$.

$$\int \frac{1}{155\sqrt{x}}\, dx = \frac{1}{155} \int \frac{1}{u} \cdot 2u\, du = \frac{2}{155} u = \frac{2}{155}\sqrt{x}.$$

Now, doing the definite integral, we have

$$\int_{1000}^{2000} \frac{1}{155\sqrt{x}}\,dx = \frac{2}{155}(\sqrt{2000} - \sqrt{1000}) \approx 0.169.$$

2. The probability that a randomly chosen individual earns more than $3000 is:

$$P(X > 3000) = \int_{3000}^{10000} \frac{1}{155\sqrt{x}}\,dx = \frac{2}{155}(\sqrt{10000} - \sqrt{3000}) \approx 0.584.$$

Example 7.28

Given a minimum income $A = 500$ and assuming a density function

$$f(x) = \begin{cases} 0 & \text{if } x < 500 \\ \frac{1}{2000}e^{-\frac{x-500}{2000}} & \text{if } x \geq 500 \end{cases}$$

1. What is the probability that a randomly chosen individual earns between 1000 and 1500?
2. What is the average income?

Solution:

1. To find the probability that a randomly chosen individual earns between 1000 and 1500:

$$P(1000 \leq X \leq 1500) = \int_{1000}^{1500} \frac{1}{2000}e^{-\frac{t-500}{2000}}\,dt.$$

This integral can be evaluated using substitution. Let $u = \frac{t-500}{2000}$, hence $du = \frac{1}{2000}dt$ and $dt = 2000\,du$. Changing the limits of integration accordingly, when $t = 1000$, $u = \frac{500}{2000} = 0.25$, and when $t = 1500$, $u = \frac{1000}{2000} = 0.5$:

$$P(1000 \leq X \leq 1500) = \int_{0.25}^{0.5} e^{-u}\,du = -(e^{-0.5} - e^{-0.25}) \approx 0.172.$$

2. For the expected value, which represents the average income:

$$E(X) = \int_{500}^{\infty} t \cdot \frac{1}{2000}e^{-\frac{t-500}{2000}}\,dt.$$

Using integration by parts, let $f = t$ and $g' = \frac{1}{2000}e^{-\frac{t-500}{2000}}\,dt$. Then

$f' = 1$ and integrating g' gives $g = -e^{-\frac{t-500}{2000}}$. So,

$$E(X) = \left[-te^{-\frac{t-500}{2000}} \right]_{500}^{\infty} + \int_{500}^{\infty} e^{-\frac{t-500}{2000}}\, dt.$$

The first term, $-te^{-\frac{t-500}{2000}}$ evaluates to 0 at the upper limit (as t approaches infinity, the exponential term goes to zero much faster than t grows), and to 500 at the lower limit ($t = 500$). In other words, $-te^{-\frac{t-500}{2000}} \to 0$ as $t \to \infty$, and at $t = 500$, it is

$$-500e^{-\frac{500-500}{2000}} = -500 \cdot 1 = -500.$$

The second term can be evaluated with another substitution (or recognizing it as the integral of an exponential function):

$$\int_{500}^{\infty} e^{-\frac{t-500}{2000}}\, dt = -2000\, e^{-\frac{t-500}{2000}} \Big|_{500}^{\infty} = -2000 \cdot (0-1) = 2000.$$

Therefore, the expected value is: $E(X) = 0 - (-500) + 2000 = 2500$.

7.5   PROBLEMS

Question 7.1 *Evaluate the definite integral:* $\int_0^2 e^{3x}\, dx$.

Question 7.2 *Use substitution to find the integral:* $\int \frac{1}{2x+1}\, dx$.

Question 7.3 *Simplify and then integrate:* $\int \frac{x^2-1}{x^3-x}\, dx$.

Question 7.4 *Compute the integrals:*

(a) $\int \frac{2x}{x^2+1}\, dx$

(b) . $\int \frac{x-1}{x^2-4x+3}\, dx$

(c) $\int \frac{4x^2+2x+2}{(x+1)(x^2+1)}\, dx$

(d) $\int x^2 e^x\, dx$

(e) $\int x\ln(x)\, dx$

(f) $\int xe^{x^2}\, dx$

(g) $\int x\sqrt{5-x^2}\, dx$

(h) $\int \frac{x}{3-x^2}\, dx$

(i) $\int \frac{x}{2-x}\, dx$

(j) $\int xe^{-3x}\, dx$

(k) $\int (3x+1)\ln x\, dx$

(l) $\int \frac{12000}{x^2+12x+36}\, dx$

(m) $\int \frac{20000}{x^2+20x+100}\, dx$

Question 7.5 *Compute the definite integrals:*

(a) $\int_0^1 \frac{3x}{x^2+4}\,dx$

(b) $\int_1^3 \frac{x-2}{x^2-3x+2}\,dx$

(c) $\int_0^1 \frac{-x^2-2x+4}{(x+1)(x^2+4)}\,dx$

(d) $\int_0^2 x^3 e^x\,dx$

(e) $\int_1^2 \ln(x) \cdot x^2\,dx$

(f) $\int_0^1 xe^{-x^2}\,dx$

(g) $\int_0^{\sqrt{5}} x\sqrt{5-x^2}\,dx$

(h) $\int_{-1}^1 \frac{x}{(3-x^2)^2}\,dx$

(i) $\int_0^1 \frac{x}{(2-x)^2}\,dx$

(j) $\int_0^\infty xe^{-2x}\,dx$

(k) $\int_1^e (3x+1)\ln x\,dx$

Question 7.6 *A company purchases a new vehicle for $50,000 and estimates that the value of the vehicle decreases at a rate given by the depreciation function $D(t) = 5,000te^{-0.2t}$ in dollars per year, where t represents the time in years. Calculate the total value of the vehicle after t years.*

Question 7.7 *A company invests in computer equipment for $200,000 and assumes that the value of this equipment decreases according to an exponential function $D(t) = 20,000te^{-0.05t}$ in dollars per year, where t is the time in years. Determine the remaining value of this equipment after t years.*

Question 7.8 *The distribution of monthly incomes in a country is such that, for all $x \geq 0$, the proportion of individuals earning less than x is given by:*

$$F(x) = \int_0^x \frac{1}{(t+200)^2}\,dt, \text{ for all } t \geq 0.$$

a. *Compute $F(x)$.*

b. *Calculate the probability of earning between 500 and 1500.*

c. *Calculate the probability of earning more than 2000.*

Question 7.9 *The distribution of monthly incomes in a country is such that, for all $x \geq 0$, the proportion of individuals earning less than x is given by:*

$$F(x) = \int_0^x \frac{1}{200}e^{\frac{-t}{200}}\,dt, \text{ for all } t \geq 0.$$

a. *Compute $F(x)$.*

b. *Calculate the probability of earning between 500 and 1500.*

c. *Calculate the probability of earning more than 2000.*

d. *Calculate the average income.*

CHAPTER 8

Multivariable Functions and Partial Derivatives

So far we have only been concerned with functions of one variable like $f(x) = x^2 - 3x + 7$ or $f(x) = e^x + \frac{1}{x}$, in other words functions with the explicit form $y = f(x)$. Notice that for these functions we have only one independent variable, usually denoted by an x. So if we input a value for x the function's output is a number. But often in applications we have functions that need more than one variable. For example, in economics the demand function does not depend only on the price of the commodity, but also on consumer's revenue, price of other commodities, and many other factors. We will study functions of more than one variable and their derivatives here. We will present the economic applications of the partial derivatives, the extrema of a function of two variables, and their economic applications.

Economics problems often involve optimization of functions of multiple variables subject to various constraints. In such cases, multivariable functions and partial derivatives play a crucial role in finding the optimal solutions. Furthermore, partial derivatives are also used to analyze the effect of changes in the input variables on the output variable in economic models. For example, in macroeconomics, partial derivatives are used to analyze the impact of changes in various economic factors, such as interest rates, inflation, and exchange rates, on the overall economy.

8.1 FUNCTIONS OF MORE THAN ONE VARIABLE

Sometimes we need to use functions that have more than one independent variable. Here is an example of a function that has two independent variables,

$$f(x, y) = 3x - 2y.$$

Notice that the function f takes in two input values, an x value and a y value. These are the two **independent variables**. Functions that have more than

DOI: 10.1201/9781003480235-8

one independent variable are called **multivariable functions**. In order to evaluate the function you need to be given both an x value and a y value. Suppose $x = 4$ and $y = -2$. Then the function would evaluate to

$$f(4, -2) = 3(4) - 2(-2) = 12 + 4 = 16.$$

In fact, functions can have any number of independent variables. Here is an example of a multivariable function with three independent variables,

$$f(x, y, z) = 2x^2 + xy - xz.$$

Here the three independent variables are called x, y, and z. Evaluating this multivariable function requires that we know the values of these three variables. Suppose $x = 1$, $y = -1$, and $z = 2$. The multivariable function would evaluate to

$$\begin{aligned} f(1, -1, 2) &= 2(1)^2 + (1)(-1) - (1)(2) \\ &= 2 - 1 - 2 \\ &= -1. \end{aligned}$$

When more than three independent variables are needed the custom is to use an indexed x. So if we needed four independent variables, we would call them x_1, x_2, x_3, and x_4. Here is an example of a multivariable function with four independent variables,

$$f(x_1, x_2, x_3, x_4) = 5x_1^2 x_3^2 - 2x_2 + 5x_4^3.$$

Fortunately for us, in this book we will not worry about multivariable functions that have more than two independent variables. However, all the ideas we will discuss are easily extendable to multivariable functions with more than two independent variables.

Example 8.1

Evaluate the multivariable function $f(x, y) = 2^x + y^2$ when $x = 3$ and $y = -3$.

Solution: We substitute 3 in for x and -3 in for y to get

$$\begin{aligned} f(3, -3) &= 2^3 + (-3)^2 \\ &= 8 + 9 \\ &= 17. \end{aligned}$$

Example 8.2

Evaluate the multivariable function $f(x, y, z) = xy^2 + yz^2 - zx^2$ when $x = -1$, $y = 0$, and $z = 1$.

Solution: We substitute -1 in for x, 0 in for y, and 1 in for z to get

$$f(-1, 0, 1) = (-1)(0)^2 + (0)(1)^2 - (1)(-1)^2$$
$$= -1.$$

The domain of definition, denoted $\mathcal{D}_f$, of the function $z = f(x, y)$ is the set of all points (x, y) for which f is defined.

Example 8.3

Evaluate the following multivariable function $f(x, y) = \sqrt{x - y}$ when $x = 1$ and $y = 3$. Then find the domain of definition $\mathcal{D}_f$.

Solution: We substitute 1 in for x, and 3 in for y to get

$$f(1, 3) = \sqrt{1 - 3}$$
$$= \sqrt{-2},$$

which is undefined. (We are not considering imaginary numbers in this book.)

If we consider this function as the elementary function $y = \sqrt{x}$, which is defined only for $x \geq 0$. This means the function $f(x, y) = \sqrt{x - y}$ is defined for all points (x, y) for which $x - y \geq 0 \Rightarrow x \geq y$. So,

$$\mathcal{D}_f = \left\{ (x, y) \in \mathbb{R} \times \mathbb{R} \, \middle| \, x \geq y \right\}.$$

The function $z = f(x, y)$ can be represented graphically in a three-dimensional coordinate system where, to each point (x, y), we associate $z = f(x, y)$. Therefore, the graph of the function f consists of all points (x, y, z) such that $z = f(x, y)$.

8.2 FIRST-ORDER PARTIAL DERIVATIVES

So far we have studied how to take the derivative of functions of one independent variable, for example, the first derivative of $f(x) = 5x^3$ is $f'(x) = 15x^2$. Notice, the function $f(x) = 5x^3$ only has one variable, the x. Now we will learn how to take derivatives of multivariable functions.

Suppose we had the function $f(x, y) = 3x - 2y$ and we wanted to take the derivative of this function. Since there are two variables, an x and a y, there are actually two first different derivatives. One derivative with respect to each variable. So we have one derivative with respect to the variable x and one derivative with respect to the variable y.

When you are asked to take the derivative of a multivariable function you need to know what variable to take the derivative with respect to. We will introduce some notation that is used. If you want to take the derivative of $f(x, y)$ with respect to x we can write

$$\frac{\partial f}{\partial x} \qquad \text{or} \qquad f_x.$$

This is sometimes called the **first-order partial derivative of f with respect to x**. If you want to take the derivative of $f(x, y)$ with respect to y we can write

$$\frac{\partial f}{\partial y} \qquad \text{or} \qquad f_y.$$

This is sometimes called the **first-order partial derivative of f with respect to y**.

Now let us see how to take the derivative of $f(x, y) = 3x - 2y$ with respect to x. This means that we will treat x as a variable. However, when we take the derivative with respect to x will just think of the y as a constant number,

$$f(x, y) = 3x - \underbrace{2y}_{\text{constant}}.$$

If we had the one variable function $g(x) = 3x - 2$ then the derivative would be $g(x) = 3$ since the derivative of the $3x$ is 3 and the derivative of the constant -2 is simply 0. similarly, the derivative of $f(x, y) = 3x - 2y$ with respect to x is given by

$$\frac{\partial f}{\partial x} = 3.$$

The derivative of $3x$ is 3 and the derivative of $-2y$ is simply 0 since we are thinking of y as a constant. When we take the derivative of $f(x, y) = 3x - 2y$ with respect to y something similar happens. We think of x as a constant and y as a variable,

$$f(x, y) = \underbrace{3x}_{\text{constant}} - 2y.$$

So the derivative with respect to y is given as

$$\frac{\partial f}{\partial y} = -2.$$

The derivative of the term $3x$ is 0 since we are thinking of x as a constant and the derivative of the term $-2y$ is -2.

Example 8.4

Given $f(x,y) = 2x^3 + 5y^4$, find the two first-order partial derivatives $\frac{\partial f}{\partial x}$ and $\frac{\partial f}{\partial y}$.

Solution: Taking the partial derivative with respect to x we consider y as a constant to get

$$\frac{\partial f}{\partial x} = 6x^2.$$

When taking the partial derivative with respect to y we consider x as a constant to get

$$\frac{\partial f}{\partial y} = 20y^3.$$

Using the other notation for derivative we could also have written

$$f_x(x,y) = 6x^2 \qquad \text{and} \qquad f_y(x,y) = 20y^3.$$

Example 8.5

Given the function $f(x,y) = -4x^5 + 7y^3$, find the two first-order partial derivatives $\frac{\partial f}{\partial x}$ and $\frac{\partial f}{\partial y}$.

Solution: Taking the partial derivative with respect to x we consider y as a constant to get

$$\frac{\partial f}{\partial x} = -20x^4.$$

When taking the partial derivative with respect to y we consider x as a constant to get

$$\frac{\partial f}{\partial y} = 21y^2.$$

Now let us see how we would take the first-order partial derivatives of a function like $f(x,y) = 5x^2y^3$. When we take the first-order partial derivative of $f(x,y)$ with respect to x we consider y to be a constant. Thus,

$$f(x,y) = \underbrace{5y^3}_{\text{constant}} x^2.$$

Since the derivative of x^2 is $2x$, we have

$$f_x(x,y) = 5y^3(2x) = 10xy^3.$$

The first-order partial derivative of $f(x,y)$ with respect to y is similar. We have

$$f(x,y) = \underbrace{5x^2}_{\text{constant}} y^3.$$

Since the derivative of y^3 is $3y^2$, we have

$$f_y(x,y) = 5x^2(3y^2) = 15x^2y^2.$$

Example 8.6

Given the function $f(x,y) = 10x^7y^6$, find the two first-order partial derivatives f_x and f_y.

Solution: When we take the partial derivative with respect to x the y is considered a constant so we get

$$f_x(x,y) = 70x^6y^6.$$

When we take the partial derivative with respect to y the x is considered a constant so we get

$$f_x(x,y) = 60x^7y^5.$$

Example 8.7

Find the two first-order partial derivatives for $f(x,y) = 3x^2y + 2y^3x - 7x + 3y - 5$.

Solution: The partial derivatives are

$$\frac{\partial f}{\partial x} = 6xy + 2y^3 - 7 \qquad \text{and} \qquad \frac{\partial f}{\partial y} = 3x^2 + 6y^2x + 3.$$

Sometimes we will need to use the chain rule. For example, suppose we want to take the partial derivative of $f(x,y) = (x^3 - y^2)^4$ with respect to x. We would need to use the chain rule to do this.

Chain Rule for Multivariable Functions:

Suppose $f(x,y) = f(u(x,y))$. Then the chain rule is given by

$$\frac{\partial f}{\partial x} = \frac{df}{du} \cdot \frac{\partial u}{\partial x} \qquad \text{and} \qquad \frac{\partial f}{\partial y} = \frac{df}{du} \cdot \frac{\partial u}{\partial y}.$$

Example 8.8

Find the two first-order partial derivatives for $f(x, y) = (x^3 - y^2)^4$.

Solution: Here we will have to use the chain rule. Letting $u = x^3 - y^2$, we have $f(u) = u^4$ so $\frac{df}{du} = 4u^3$. We also have $\frac{\partial u}{\partial x} = 3x^2$. Using the first formula for the chain rule, we have

$$\frac{\partial f}{\partial x} = \frac{df}{du} \cdot \frac{\partial u}{\partial x}$$
$$= 4u^3(3x^2)$$
$$= 4(x^3 - y^2)^3(3x^2)$$
$$= 12x^2(x^3 - y^2)^3.$$

Again, letting $u = x^3 - y^2$, we have $f(u) = u^4$ and $\frac{df}{du} = 4u^3$. We also have $\frac{\partial u}{\partial y} = -2y$. Using the second formula for the chain rule, we have

$$\frac{\partial f}{\partial y} = \frac{df}{du} \cdot \frac{\partial u}{\partial y}$$
$$= 4u^3(-2y)$$
$$= 4(x^3 - y^2)^3(-2y)$$
$$= -8y(x^3 - y^2)^3.$$

Example 8.9

Find the two first-order derivatives for $f(x, y) = e^{x^2 - y^2}$.

Solution: We write $f(u) = e^u$ where $u = x^2 - y^2$. Since $\frac{df}{du} = e^u$ and $\frac{\partial u}{\partial x} = 2x$ the partial derivative of f with respect to x is

$$\frac{\partial f}{\partial x} = \frac{df}{du} \cdot \frac{\partial u}{\partial x}$$
$$= e^u(2x)$$
$$= 2xe^{x^2 - y^2}.$$

Similarly, we find the partial derivative of f with respect to y is

$$\frac{\partial f}{\partial y} = \frac{df}{du} \cdot \frac{\partial u}{\partial y}$$
$$= e^u(-2y)$$
$$= -2ye^{x^2 - y^2}.$$

Example 8.10

Find the two first-order partial derivatives of $f(x, y) = 3 \ln(2x^5 y^4)$.

Solution: We let $f(u) = 3 \ln(u)$ and let $u = 2x^5 y^4$. Thus, $\frac{df}{du} = \frac{3}{u}$ and $\frac{\partial u}{\partial x} = 10x^4 y^4$, so the partial derivative of f with respect to x is

$$\frac{\partial f}{\partial x} = \frac{df}{du} \cdot \frac{\partial u}{\partial x}$$
$$= \frac{3}{u}(10x^4 y^4)$$
$$= \frac{30x^4 y^4}{2x^5 y^4}$$
$$= \frac{15}{x}.$$

Similarly, the partial derivative of f with respect to y is

$$\frac{\partial f}{\partial y} = \frac{df}{du} \cdot \frac{\partial u}{\partial y}$$
$$= \frac{3}{u}(8x^5 y^3)$$
$$= \frac{24x^5 y^3}{2x^5 y^4}$$
$$= \frac{12}{y}.$$

Example 8.11

Find the two first-order partial derivatives of $f(x, y) = \dfrac{1}{\sqrt{x^3 + y^3}}$.

Solution: We have $f(u) = \frac{1}{\sqrt{u}} = u^{-\frac{1}{2}}$ where $u = x^3 + y^3$. Clearly $\frac{df}{du} = -\frac{1}{2}u^{-\frac{3}{2}}$ and $\frac{\partial u}{\partial x} = 3x^2$. This gives us the partial derivative of f with respect to x,

$$\frac{\partial f}{\partial x} = \frac{df}{du} \cdot \frac{\partial u}{\partial x}$$
$$= -\frac{1}{2}u^{-\frac{3}{2}}(3x^2)$$
$$= \frac{-3x^2}{2(x^3 + y^3)^{\frac{3}{2}}}.$$

Similarly, the partial derivative of f with respect to y is

$$\frac{\partial f}{\partial y} = \frac{df}{du} \cdot \frac{\partial u}{\partial y}$$

$$= -\frac{1}{2}u^{-\frac{3}{2}}(3y^2)$$

$$= \frac{-3y^2}{2(x^3 + y^3)^{\frac{3}{2}}}.$$

Suppose we could write f as the product of two multivariable functions u and v. To find the two first-order partial derivatives, we would need to use the mutivariable version of the product rule.

Product Rule for Multivariable Functions:

Suppose $f(x,y) = u(x,y) \cdot v(x,y)$. Then the product rule is given by

$$\frac{\partial f}{\partial x} = \frac{\partial u}{\partial x} \cdot v + u \cdot \frac{\partial v}{\partial x} \qquad \text{and} \qquad \frac{\partial f}{\partial y} = \frac{\partial u}{\partial y} \cdot v + u \cdot \frac{\partial v}{\partial y}.$$

Example 8.12

Find the two first-order derivatives for $f(x,y) = (4x^3 - 5y^2)(3x^2y + xy^3)$.

Solution: Let $f(x,y) = u(x,y) \cdot v(x,y)$ with $u(x,y) = 4x^3 - 5y^2$ and $v(x,y) = 3x^2y + xy^3$. Thus,

$$\frac{\partial u}{\partial x} = 12x^2, \qquad \frac{\partial v}{\partial x} = 6xy + y^3,$$

$$\frac{\partial u}{\partial y} = -10y, \qquad \frac{\partial v}{\partial y} = 3x^2 + 3xy^2.$$

The first-order partial derivative of f with respect to x is given by

$$\frac{\partial f}{\partial x} = \frac{\partial u}{\partial x} \cdot v + u \cdot \frac{\partial v}{\partial x}$$

$$= (12x^2)(3x^2y + xy^3) + (4x^3 - 5y^2)(6xy + y^3),$$

and the first-order partial derivative of f with respect to y is given by

$$\frac{\partial f}{\partial y} = \frac{\partial u}{\partial y} \cdot v + u \cdot \frac{\partial v}{\partial y}$$

$$= (-10y)(3x^2y + xy^3) + (4x^3 - 5y^2)(3x^2 + 3xy^2).$$

Of course, we could simplify the final answers in the above example, but that adds little to understanding the product rule so we will not do that here.

Example 8.13

Find the two first-order derivatives for $f(x,y) = \ln(xy^2)e^{x-y^3+2}$.

Solution: First we write $f(x,y) = u(x,y) \cdot v(x,y)$ where $u(x,y) = \ln(xy^2)$ and $v(x,y) = e^{x-y^3+2}$. It is easy to see

$$\frac{\partial u}{\partial x} = \frac{y^2}{xy^2} = \frac{1}{x}, \qquad \frac{\partial v}{\partial x} = e^{x-y^3+2},$$

$$\frac{\partial u}{\partial y} = \frac{2xy}{xy^2} = \frac{2}{y}, \qquad \frac{\partial v}{\partial y} = -3y^2 e^{x-y^3+2}.$$

The first-order partial derivative of f with respect to x is given by

$$\frac{\partial f}{\partial x} = \frac{\partial u}{\partial x} \cdot v + u \cdot \frac{\partial v}{\partial x}$$

$$= \frac{1}{x} e^{x-y^3+2} + \ln(xy^2)e^{x-y^3+2},$$

and the first-order partial derivative of f with respect to y is given by

$$\frac{\partial f}{\partial y} = \frac{\partial u}{\partial y} \cdot v + u \cdot \frac{\partial v}{\partial y}$$

$$= \frac{2}{y} e^{x-y^3+2} + \ln(xy^2)(-3y^2 e^{x-y^3+2}).$$

Suppose we could write f as the quotient of two multivariable functions u and v. To find the two first-order partial derivatives, we would need to use the mutivariable version of the quotient rule.

Quotient Rule for Multivariable Functions:

Suppose $f(x,y) = \dfrac{u(x,y)}{v(x,y)}$. Then the quotient rule is given by

$$\frac{\partial f}{\partial x} = \frac{v \cdot \frac{\partial u}{\partial x} - u \cdot \frac{\partial v}{\partial x}}{v^2} \qquad \text{and} \qquad \frac{\partial f}{\partial y} = \frac{v \cdot \frac{\partial u}{\partial y} - u \cdot \frac{\partial v}{\partial y}}{v^2}.$$

Example 8.14

Find the two first-order derivatives for $f(x,y) = \dfrac{4x^3 - 5y^2}{3x^2y + xy^3}$.

Solution: We first we write $f(x,y) = \dfrac{u(x,y)}{v(x,y)}$ where

$$u(x,y) = 4x^3 - 5y^2 \qquad \text{and} \qquad v(x,y) = 3x^2y + xy^3.$$

As in the last example, it is easy to see

$$\frac{\partial u}{\partial x} = 12x^3, \qquad \frac{\partial v}{\partial x} = 6xy + y^3,$$

$$\frac{\partial u}{\partial y} = -10y, \qquad \frac{\partial v}{\partial y} = 3x^2 + 3xy^2.$$

The first-order partial derivative of f with respect to x is given by

$$\frac{\partial f}{\partial x} = \frac{v \cdot \frac{\partial u}{\partial x} - u \cdot \frac{\partial v}{\partial x}}{v^2}$$

$$= \frac{\left(3x^2y + xy^3\right)\left(12x^3\right) - \left(4x^3 - 5y^2\right)\left(6xy + y^3\right)}{\left(3x^2y + xy^3\right)^2}$$

and the first-order partial derivative of f with respect to y is given by

$$\frac{\partial f}{\partial y} = \frac{v \cdot \frac{\partial u}{\partial y} - u \cdot \frac{\partial v}{\partial y}}{v^2}$$

$$= \frac{\left(3x^2y + xy^3\right)\left(-10y\right) - \left(4x^3 - 5y^2\right)\left(3x^2 + 3xy^2\right)}{\left(3x^2y + xy^3\right)^2}.$$

We will not attempt to simplify these expressions.

Example 8.15

Find the two first-order derivatives for $f(x,y) = \dfrac{e^{x^3 - 5y^2}}{xy^2 - y^3}$.

Solution: Again, we first we write $f(x,y) = \dfrac{u(x,y)}{v(x,y)}$ where $u(x,y) = e^{x^3 - 5y^2}$ and $v(x,y) = xy^2 - y^3$. As in the last example, it is easy to see

$$\frac{\partial u}{\partial x} = 3x^2 e^{x^3 - 5y^2}, \qquad \frac{\partial v}{\partial x} = y^2,$$

$$\frac{\partial u}{\partial y} = -10y e^{x^3 - 5y^2}, \qquad \frac{\partial v}{\partial y} = 2xy - 3y^2.$$

The first-order partial derivative of f with respect to x is given by

$$\frac{\partial f}{\partial x} = \frac{v \cdot \frac{\partial u}{\partial x} - u \cdot \frac{\partial v}{\partial x}}{v^2}$$

$$= \frac{\left(xy^2 - y^3\right)\left(3x^2 e^{x^3 - 5y^2}\right) - y^2 e^{x^3 - 5y^2}}{\left(xy^2 - y^3\right)^2}$$

$$= \frac{e^{x^3 - 5y^2}\left[3x^2\left(xy^2 - y^3\right) - y^2\right]}{\left(xy^2 - y^3\right)^2}$$

and the first-order partial derivative of f with respect to y is given by

$$\frac{\partial f}{\partial y} = \frac{v \cdot \frac{\partial u}{\partial y} - u \cdot \frac{\partial v}{\partial y}}{v^2}$$

$$= \frac{\left(xy^2 - y^3\right)\left(-10y e^{x^3 - 5y^2}\right) - \left(2xy - 3y^2\right) e^{x^3 - 5y^2}}{\left(xy^2 - y^3\right)^2}$$

$$= \frac{e^{x^3 - 5y^2}\left[-10y\left(xy^2 - y^3\right) - \left(2xy - 3y^2\right)\right]}{\left(xy^2 - y^3\right)^2}.$$

8.3 SECOND-ORDER PARTIAL DERIVATIVES

Of course, we can also take second derivatives of multivariable functions just like we took second derivatives of one-variable functions. Recall, the second derivative is simply the derivative of the first derivative. But when we took the first derivative of the two variable function $f(x, y)$ we had two choices; we could either take the derivative of f with respect to x to get $\frac{\partial f}{\partial x}$, or we could take the derivative of f with respect to y to get $\frac{\partial f}{\partial y}$.

If we have the first-order partial derivative $\frac{\partial f}{\partial x}$, we have two choices for the second derivative, we can either take the derivative of $\frac{\partial f}{\partial x}$ with respect to x or with respect to y. These would be written as

$$\frac{\partial}{\partial x}\left(\frac{\partial f}{\partial x}\right) = \frac{\partial^2 f}{\partial x \partial x} = \frac{\partial^2 f}{\partial x^2} \qquad \text{and} \qquad \frac{\partial}{\partial y}\left(\frac{\partial f}{\partial x}\right) = \frac{\partial^2 f}{\partial y \partial x}.$$

If we had the first-order partial derivative $\frac{\partial f}{\partial y}$ we again have two choices for the second derivative, we can either take the derivative of $\frac{\partial f}{\partial y}$ with respect to

x or with respect to y. These would be written as

$$\frac{\partial}{\partial x}\left(\frac{\partial f}{\partial y}\right) = \frac{\partial^2 f}{\partial x \partial y} \quad \text{and} \quad \frac{\partial}{\partial y}\left(\frac{\partial f}{\partial y}\right) = \frac{\partial^2 f}{\partial y \partial y} = \frac{\partial^2 f}{\partial y^2}.$$

Thus there are four **second-order partial derivatives of** f which are usually written as

$$\frac{\partial^2 f}{\partial x^2}, \quad \frac{\partial^2 f}{\partial y \partial x}, \quad \frac{\partial^2 f}{\partial x \partial y}, \quad \frac{\partial^2 f}{\partial y^2}.$$

As long as $\frac{\partial f}{\partial x}$ and $\frac{\partial f}{\partial x}$ are continuous, something really nice happens. It turns out that

$$\frac{\partial^2 f}{\partial y \partial x} = \frac{\partial^2 f}{\partial x \partial y}.$$

This is called the equality of mixed-partials or known as **Young's Theorem**. In general, if you know how to take first-order partial derivatives then you can take second-order partial derivatives.

Example 8.16

Find the second-order partial derivatives of $f(x, y) = 4x^7 y^5$.

Solution: First we find the first-order derivative of f with respect to x,

$$\frac{\partial f}{\partial x} = 28x^6 y^5,$$

which we use to find the second-order partial derivatives

$$\frac{\partial^2 f}{\partial x^2} = 168x^5 y^5 \quad \text{and} \quad \frac{\partial^2 f}{\partial y \partial x} = 140x^6 y^4.$$

Next, we find the first-order derivative of f with respect to y,

$$\frac{\partial f}{\partial y} = 20x^7 y^4,$$

which we use to find the second-order partial derivatives

$$\frac{\partial^2 f}{\partial x \partial y} = 140x^6 y^4 \quad \text{and} \quad \frac{\partial^2 f}{\partial y^2} = 80x^7 y^3.$$

Notice the equality of the mixed partials.

Taking second derivatives sometimes requires using the chain rule, the product rule, or the quotient rule.

Example 8.17

Given $f(x, y) = \ln\left(x^2y - x\right)$, find $\dfrac{\partial^2 f}{\partial x^2}$.

Solution: First we must find the first-order derivative of f with respect to x, which requires using the chain rule. This gives

$$\frac{\partial f}{\partial x} = \frac{2xy - 1}{x^2y - x}.$$

Taking the second-order partial derivative requires using the quotient rule. Here we let $u(x, y) = 2xy - 1$ and $v(x, y) = x^2y - x$, so we have

$$\frac{\partial^2 f}{\partial x^2} = \frac{v \cdot \frac{\partial u}{\partial x} - u \cdot \frac{\partial v}{\partial x}}{v^2}$$

$$= \frac{\left(x^2y - x\right)(2y) - (2xy - 1)(2xy - 1)}{\left(x^2y - x\right)^2}$$

where we will forego simplification.

Example 8.18

Given $f(x, y) = \ln(x^2y - x)$, find $\dfrac{\partial^2 f}{\partial y \partial x}$.

Solution: As in the last example, we first find

$$\frac{\partial f}{\partial x} = \frac{2xy - 1}{x^2y - x},$$

which is then used to find the second order partial derivative using the quotient rule. Again, letting $u(x, y) = 2xy - 1$ and $v(x, y) = x^2y - x$, we have

$$\frac{\partial^2 f}{\partial y \partial x} = \frac{v \cdot \frac{\partial u}{\partial y} - u \cdot \frac{\partial v}{\partial y}}{v^2}$$

$$= \frac{\left(x^2y - x\right)(2x) - (2xy - 1)\left(x^2\right)}{\left(x^2y - x\right)^2}$$

where we again forego simplification.

Example 8.19

Given $f(x, y) = \ln(x^2 y - x)$, find $\dfrac{\partial^2 f}{\partial y^2}$.

Solution: The first-order derivative of f with respect to y requires using the chain rule. This gives

$$\frac{\partial f}{\partial y} = \frac{x^2}{x^2 y - x}.$$

Taking the second-order partial derivative requires using the quotient rule. Here, we let $u(x, y) = x^2$ and $v(x, y) = x^2 y - x$, so we have

$$\frac{\partial^2 f}{\partial y^2} = \frac{v \cdot \frac{\partial u}{\partial y} - u \cdot \frac{\partial v}{\partial y}}{v^2}$$

$$= \frac{(x^2 y - x)(0) - (x^2)(x^2)}{(x^2 y - x)^2}$$

$$= \frac{-x^4}{(x^2 y - x)^2}.$$

8.4 APPLICATIONS OF PARTIAL DERIVATIVES

Total Differentials and Incremental Change

For a function of one independent variable $y = f(x)$, we define the derivative as $\frac{dy}{dx} = f'(x)$. This is used to define what we call the **total differential** of y as

$$dy = f'(x)dx,$$

which can also be written as

$$dy = \frac{dy}{dx}dx.$$

Essentially, it looks like we multiplied both sides of $\frac{dy}{dx} = f'(x)$ by the "infinitesimal" dx. This formulation is important because it ties an infinitesimally small change in x, denoted by dx, to the corresponding infinitesimal change in y, denoted by dy, via the derivative $f'(x)$. We can use this formula to estimate the change in y given a small change in x. It we let Δx represent the small change in x and Δy represent the small change in y, we have

$$\Delta y \approx f'(x)\,\Delta x.$$

Example 8.20

Consider a company that produces and sells coffee. The company's analysis of past data suggests that revenue from coffee sales can be represented by a function of the price per unit, $y = R(x) = 100x - 0.5x^2$, where y is the revenue and x is the price per unit of coffee. The company wants to understand how a small change in the price of coffee will affect its revenue.

Solution: Here, x represents the current price per unit, and Δx represents a small increase in the price. The derivative $\frac{dy}{dx}$ would then represent the rate at which revenue changes with respect to a change in price. Thus, the first derivative of this function, $R'(x) = 100 - x$, tells us how revenue changes for each dollar increase in price. Using the total differential, we can calculate the approximate change in revenue Δy for a small increase in price Δx,

$$\Delta y \approx R'(x)\Delta x = (100 - x)\,\Delta x.$$

Suppose the current price per unit is \$50 and the company is considering a \$1 increase, which means $\Delta x = 1$. Thus,

$$\Delta y \approx (100 - 50) \cdot 1 = 50.$$

This means that for a \$1 increase in the price per unit, in other words, for a price increase from \$50 to \$51, the company can expect approximately a \$50 increase in total revenue.

Example 8.21

Consider a technology company that specializes in producing a popular line of smartphones. The company wishes to understand how minor adjustments to the smartphone's selling price will affect their overall revenue. Based on historical data and market analysis, the relationship between the selling price x and the revenue y is best described by the function $y = 150x - x^2$.

Solution: In this scenario, x denotes the current selling price per smartphone, and Δx symbolizes a slight increase in this price and the derivative $\frac{dy}{dx}$ signifies the rate at which revenue evolves in response to changes in the selling price. The first derivative of this revenue function, $\frac{dy}{dx} = 150 - 2x$, indicates how revenue shifts for each dollar increase in the selling price. To calculate the estimated change in revenue Δy for a minor price increase Δx, we utilize the total differential,

$$\Delta y \approx \frac{dy}{dx}\Delta x = (150 - 2x)\,\Delta x.$$

Suppose the current selling price per smartphone is $75 and the company is contemplating a $2 increase. Thus,

$$\Delta y \approx (150 - 2 \cdot 75) \cdot 2 = (150 - 150) \cdot 2 = 0.$$

This outcome indicates that for a $2 increase in the selling price per unit from $75, the company would not see any change in revenue, assuming other factors remain constant.

Similarly, for a function of two independent variables $z = f(x, y)$, we define total differential of z by:

$$dz = \left(\frac{\partial f}{\partial x}\right)dx + \left(\frac{\partial f}{\partial y}\right)dy.$$

How is this useful to us? Suppose we have a small change in the value of x and y, which we will denote by Δx and Δy, and we want to find the associated change in z, denoted by Δz. This would be given by

$$f(x + \Delta x, y + \Delta y) = f(x, y) + \Delta z$$
$$\implies \Delta z = f(x + \Delta x, y + \Delta y) - f(x, y).$$

However, when Δx and Δy are very small then we can assume that $dx \approx \Delta x$ and $dy \approx \Delta y$, which means that we can approximate Δz using the total differential; that is,

$$\Delta z \approx \left(\frac{\partial f}{\partial x}\right)\Delta x + \left(\frac{\partial f}{\partial y}\right)\Delta y,$$

where Δz represents the approximate change in z when x and y change.

Example 8.22

1. Find the total differential of z for $z = 3x^2 y^3$.
2. Use the total differential to find the approximate change in z if x increases by 4% and y deceases by 2%.

Solution:

1. The total differential of $z = 3x^2 y^3$ is given by $\Delta z = \left(\frac{\partial f}{\partial x}\right)\Delta x + \left(\frac{\partial f}{\partial y}\right)\Delta y$. Since $\frac{\partial f}{\partial x} = 6xy^3$ and $\frac{\partial f}{\partial y} = 9x^2 y^2$, we have

$$\Delta z \approx 6xy^3\,\Delta x + 9x^2 y^2\,\Delta y.$$

2. A 4% increase in the value of x is given by $\Delta x = \frac{4}{100}x$ and a 2% decrease in the value of y is given by $\Delta y = -\frac{2}{100}y$, therefore,

$$\Delta z \approx 6xy^3 \left(\frac{4}{100}x\right) + 9x^2y^2 \left(\frac{-2}{100}y\right)$$

$$= 6x^2y^3 \frac{4}{100} - 9x^2y^3 \frac{2}{100}$$

$$= 3x^2y^3 \left(\frac{8}{100} - \frac{6}{100}\right)$$

$$= z\frac{2}{100}.$$

Therefore, there will be a 2% increase in z if x increases by 4% and y decreases by 2%.

The total differential provides a first-order approximation of how a function changes in response to small changes in its inputs. This concept is not only fundamental in theoretical mathematics but is also extensively used in various fields to analyze and predict the behavior of complex systems. Consider, for example, a situation in economics where z represents the total cost and x and y represent quantities of two different inputs used in production. If the prices of these inputs change slightly, the total differential can estimate how the total cost z will be affected without needing to re-evaluate the entire cost function for every small change.

Consider a company that is planning its budget for the next fiscal year. The company's revenue, R, depends on multiple factors such as market demand, D, and the price of its products, P. Thus, we can express the revenue as a function of these variables, $R = f(D, P)$. To forecast revenue for the next year, the company needs to account for changes in both market demand and product prices. Let us denote the change in market demand by ΔD and the change in product price by ΔP.

The total change in revenue, ΔR, due to changes in market demand and price can be approximated using a method analogous to the previous example;

$$\Delta R \approx \left(\frac{\partial R}{\partial D}\right)\Delta D + \left(\frac{\partial R}{\partial P}\right)\Delta P.$$

Suppose the initial analysis suggests that if market demand increases by 5% ($\Delta D = 5\%$) and the company decides to increase the price of its products by 2% ($\Delta P = 2\%$), how this will affect the company revenue? By applying the formula above, the company can approximate the change in revenue based on the sensitivity of revenue to both demand and price changes, which are expressed by the partial derivatives $\frac{\partial R}{\partial D}$ and $\frac{\partial R}{\partial P}$, respectively. This approach allows the company to make informed decisions by estimating the impact of various external and internal factors on its revenue. Let us examine this example in more details.

Example 8.23

A company produces a single product. The revenue of this company, R, depends on two variables: market demand D in units and the price of its product P in dollars. The relationship between revenue, market demand, and product price is given by the function

$$R(D, P) = D \cdot P.$$

The current market demand D is 10,000 units and the current price P is \$50 per unit. Find the current revenue and the predicted revenue next year if market demand is expected to increase by 5% and the company plans to raise prices by 2%.

Solution: The current revenue R can be calculated to be

$$R = 10{,}000 \cdot 50 = \$500{,}000.$$

Since the company expects that market demand will increase by 5% and plans to increase the price of its product by 2%. Therefore, we have

$$\Delta D = 5\% \times 10{,}000 = 500 \text{ (units)},$$
$$\Delta P = 2\% \times 50 = \$1 \text{ (per unit)}.$$

The total differential formula to approximate the change in revenue ΔR is

$$\Delta R \approx \left(\frac{\partial R}{\partial D}\right) \Delta D + \left(\frac{\partial R}{\partial P}\right) \Delta P.$$

Therefore, to find the total change in revenue ΔR using the total differential, we first determine the partial derivatives of R with respect to D and P:

$$\frac{\partial R}{\partial D} = P \quad \text{and} \quad \frac{\partial R}{\partial P} = D.$$

Substituting into the total differential formula for ΔR gives us

$$\Delta R \approx P \, \Delta D + D \, \Delta P.$$

This in turn gives us

$$\Delta R \approx (50) \cdot 500 + (10{,}000) \cdot 1 = 35{,}000.$$

Thus, the approximate increase in revenue, given the 5% increase in demand and 2% increase in price, is \$35,000. This gives a predicted revenue of $500{,}000 + 35{,}000 = 535{,}000$ dollars.

Example 8.24

Consider a company that manufactures electronic gadgets with the total cost of production C depending on two variables: the number of units produced Q and the cost of raw materials per unit M. This relationship is given by

$$C(Q, M) = aQ + bMQ + c,$$

where

- a represents the cost per unit of production,
- b represents the variable cost coefficient,
- c represents the fixed costs unrelated to the level of production.

Suppose the current production level Q is 1,000 units and the cost of raw materials per unit M is \$20, with fixed cost per unit a of \$5, variable cost coefficient b of \$2, and total fixed costs c of \$10,000. The company anticipates an increase in production by 100 units ($\Delta Q = 100$) and an increase in the cost of raw materials by \$1 ($\Delta M = 1$). Find the anticipated change in cost of production.

Solution: Given $C(Q, M) = aQ + bMQ + c$, the partial derivatives of C with respect to Q and M are

$$\frac{\partial C}{\partial Q} = a + bM \qquad \text{and} \qquad \frac{\partial C}{\partial M} = bQ.$$

We also have the given values

- $Q = 1000$,
- $M = \$20$,
- $\Delta Q = 100$,

- $\Delta M = \$1$,

- $a = \$5$,

- $b = \$2$,

- $c = \$10,000$.

Substituting the given values we get

$$\frac{\partial C}{\partial Q} = 5 + 2 \times 20 = 45, \qquad \text{and} \qquad \frac{\partial C}{\partial M} = 2 \times 1000 = 2000.$$

We can then estimate the total change in production costs ΔC using the total differential,

$$\Delta C \approx \left(\frac{\partial C}{\partial Q}\right) \Delta Q + \left(\frac{\partial C}{\partial M}\right) \Delta M$$
$$= 45(100) + 2000(1) = 6500.$$

Therefore, the estimated change in total costs, ΔC, with an increase in production by 100 units and an increase in the cost of raw materials by \$1 per unit, is \$6500.

Example 8.25

Consider a company that sells a software subscription service. The revenue R depends on the subscription price P and the number of subscribers S, modeled by $R(P, S) = P \cdot S$. The current subscription price P is \$30 per month and the current number of subscribers S is 2000. The company predicts that increasing the subscription price by \$5 might result in losing 100 subscribers. Calculate the estimated change in revenue.

Solution: We have $\Delta P = 5$ and $\Delta S = -100$. Given $R(P, S) = P \cdot S$, the partial derivatives of the revenue function with respect to P and S are

$$\frac{\partial R}{\partial P} = S \quad \text{and} \quad \frac{\partial R}{\partial S} = P.$$

Using the total differential to estimate the change in revenue ΔR, we get

$$\Delta R \approx \left(\frac{\partial R}{\partial P}\right)\Delta P + \left(\frac{\partial R}{\partial S}\right)\Delta S.$$

Substituting the partial derivatives into this equation, we have

$$\Delta R \approx S\Delta P + P\Delta S$$
$$= 2000(5) + 30(-100) = 7000.$$

Therefore, the estimated change in revenue, based on the total differential, is an increase of \$7,000.

Marginal Analysis for Multivariable Functions

Multivariable functions often appear in business applications. One way that this occurs is when a company makes and sells more than one kind of product. Suppose a company makes two different types of widgets, type X and type Y. Each week the company sells x type X widgets and y type Y widgets. The company has the following cost and revenue functions:

$$C(x, y) = 7x + 9y + 18{,}000,$$
$$R(x, y) = 50x + 70y + 0.07xy - 0.04x^2 - 0.04y^2.$$

Suppose we wanted to know what the company's cost and revenue were when it sold 800 type X widgets and 1000 type Y widgets. We would simply evaluate the functions at $x = 800$ and $y = 1000$,

$$C(800, 1000) = 32{,}600$$
$$R(800, 1000) = 100{,}400.$$

We can also perform marginal analysis on multivariable functions. Let us start by looking at the cost function. We are now producing two types of widgets, type X and type Y. We have two marginal cost functions, the **marginal cost function for producing widget X** (while production of widget Y is held constant) and the **marginal cost function for producing widget Y** (when production of widget X is held constant.) The marginal cost function for producing widget X is given by the partial derivative of the cost function with respect to x,

$$C_x(x,y) = 7.$$

Recall that C_x is simply another way of writing $\frac{\partial C}{\partial x}$. What is the marginal cost for widget X when production is $x = 800$ and $y = 1000$? We simply evaluate C_x at the relevant values,

$$C_x(800, 1000) = 7.$$

What this means is that the cost of producing one additional widget X when production of type X widget is $x = 800$ and type Y widget is $y = 1000$. Another way of saying this is that the cost of producing the 801st widget X when production of widget Y is held constant at 1000 is 7 dollars. Similarly, the marginal cost function for producing widget Y is given by the partial derivative of the cost function with respect to y,

$$C_y(x,y) = 9.$$

The marginal cost for widget Y when production is $x = 800$ and $y = 1000$ is

$$C_y(800, 1000) = 9.$$

This means that the cost of producing the 1001st widget Y when production of widget X is held constant at 800 is 9 dollars. Marginal revenue is of course handled exactly the same way.

Example 8.26

Estimate the revenue from selling the 801st widget X when widget Y sales are constant at 1000. Then estimate the revenue of selling the 1001st widget Y when widget X sales are constant at 800. The revenue function is given by $R(x,y) = 50x + 70y + 0.07xy - 0.04x^2 - 0.04y^2$.

Solution: We first find the marginal revenue function for widget X while holding widget Y production constant,

$$R_x(x,y) = 50 + 0.07y - 0.08x.$$

To estimate the revenue from selling the 801st widget X when widget Y sales are constant at 1000 we evaluate this at $x = 800$ and $y = 1000$,

$$R_x(800, 1000) = 50 + 0.07(1000) - 0.08(800) = 56.$$

Next, we find the marginal revenue function for widget Y while holding widget X production constant,

$$R_y(x, y) = 70 + 0.07x - 0.08y.$$

To estimate the revenue of selling the 1001st widget Y when widget X sales are constant at 800 we evaluate this at $x = 800$ and $y = 1000$,

$$R_y(800, 1000) = 70 + 0.07(800) - 0.08(1000) = 46.$$

A company that makes two products may have a different price-demand function for each product. Suppose a company makes two items, X and Y. Item X has a price-demand function

$$p_X = 200 - 4x + y$$

and item Y had price demand function

$$p_Y = 400 + 5x - 2y,$$

where x is the number of item X produced and y is the number of item Y produced. Note, here the subscripts in p_X and p_Y refer to the type of widget the price-demand function refers to, not to a partial derivative variable. The price-demand functions for these two items can be used to find the revenue function for this company. The revenue function is given by

$$\begin{aligned} R(x, y) &= p_X \cdot x + p_Y \cdot y \\ &= (200 - 4x + y)x + (400 + 5x - 2y)y \\ &= 200x - 4x^2 + xy + 400y + 5xy - 2y^2 \\ &= 200x + 400y + 6xy - 4x^2 - 2y^2. \end{aligned}$$

Example 8.27

Suppose a company makes two items, X and Y. Item X has a price-demand function $p_X = 200 - 4x + y$ and item Y had price demand function $p_Y = 400 + 5x - 2y$. The cost function is given by $C(x, y) = 500 + 50x + 100y$. Estimate the expected profit from selling the 501st item of X when sales of item Y are held constant at 700.

Solution: First we must use the two price-demand functions to find the revenue function. This was done above,

$$R(x, y) = 200x + 400y + 6xy - 4x^2 - 2y^2.$$

Next, we use this and the cost function to find the profit function,

$$P(x, y) = R(x, y) - C(x, y)$$
$$= (200x + 400y + 6xy - 4x^2 - 2y^2) - (500 + 50x + 100y)$$
$$= -500 + 150x + 300y + 6xy - 4x^2 - 2y^2.$$

To estimate the expected profit from selling the 501st item X when sales of item Y are held constant at 700 we need the marginal profit function for item X,

$$P_x(x, y) = 150 + 6y - 8x.$$

We then evaluate this at $x = 500$ and $y = 700$,

$$P_x(500, 700) = 150 + 6(700) - 8(500) = 350.$$

Thus, the expected profit from selling the 501st item of X when sales of item Y are held constant at 700 is $350.

Marginal productivity

The production of any good generally requires the use of different **production factors** such as labor, capital, land, materials, or machines. In general, production of anything will require more than one production factor. Let us assume a production function $z = f(x, y)$ where a quantity z of a good is made using the quantities x and y of two production factors. The partial derivative of z with respect to x gives us the **marginal productivity** of x, and the partial derivative of z with respect to y which gives us the marginal productivity of y.

Example 8.28

If the production function of a good is given by: $z = 5xy - 2x^3 + y^3$.

Solution: The marginal productivity of x is equal to

$$\frac{\partial z}{\partial x} = 5y - 6x^2$$

and the marginal productivity of y is equal to

$$\frac{\partial z}{\partial y} = 5x + 3y^2.$$

Example 8.29

If the production function of a good is given by: $f(K, L) = K^3 L^2$, where K and L represent the capital and the labor quantities. This type of production function is called the Cobb-Douglas production function that we will presented in detail in Chapter 9.

Solution: The marginal productivity (MP), of capital is equal to

$$MP_K = \frac{\partial f}{\partial K} = 3K^2 L^2$$

and the marginal productivity of labor is equal to

$$MP_L = \frac{\partial f}{\partial L} = 2K^3 L.$$

Minima and maxima of a function of two variables

A function of two variables $z = f(x, y)$ has a maximum at the point $\left(a, b, f(a, b)\right)$ if $f(a, b)$ has a value greater than $f(x, y)$ for all other points (x, y) in a neighborhood of $x = a$ and $y = b$; that is, for all points (x, y) that are close to (a, b). Here is an example of a function that has a local maximum.

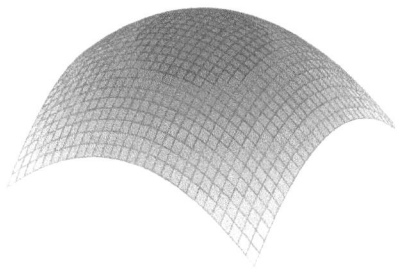

Likewise, $f(x, y)$ has a minimum at the point $\left(a, b, f(a, b)\right)$ if $f(a, b)$ has a lower value than $f(x, y)$ for all other points (x, y) in a neighborhood of $x = a$ and $y = b$. Here is an example of a function with a local minimum.

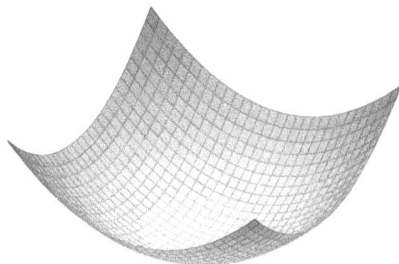

Similarly to what we have seen with function of one independent variable, there is a horizontal tangent plane at the maximum or minimum points $(a, b, f(a, b))$. This tangent plane is generated by the two tangent lines determined by $\frac{\partial f}{\partial x}$ and $\frac{\partial f}{\partial y}$. In other words, for (x, y) to be a minimum or a maximum to the function f, it is necessary that the following two equations are satisfied: $\frac{\partial f}{\partial x} = 0$ and $\frac{\partial f}{\partial y} = 0$. Solving these two equations for the points x and y gives us the critical values of the function. Note, we have two equations and two unknowns, but as long as the functions are not too difficult we can generally find the critical points. But we should note that not every critical point results in a maximum or a minimum. Another possibility is a saddle point. Here is an example of saddle point behavior.

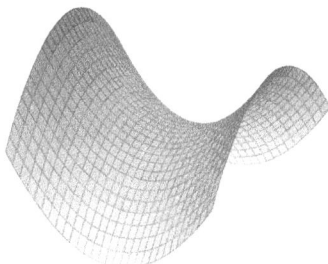

We need a way to determine if our critical point is a maximum, a minimum, or a saddle point. Similar to the second derivative test, we need a condition that is sufficient for the existence of a minimum or a maximum. This condition is

$$\frac{\partial^2 f}{\partial x^2} \times \frac{\partial^2 f}{\partial y^2} - \left(\frac{\partial^2 f}{\partial x \partial y} \right)^2 > 0.$$

We shall not prove this condition here. If this condition is satisfied for our critical point(s), then we know it is either a maximum or a minimum, but we still do not know which. If $\frac{\partial^2 f}{\partial x^2} > 0$, then the critical point is a minimum, and if $\frac{\partial^2 f}{\partial x^2} < 0$, then the critical point is a maximum. Again, we will not prove this here. However, if

$$\frac{\partial^2 f}{\partial x^2} \times \frac{\partial^2 f}{\partial y^2} - \left(\frac{\partial^2 f}{\partial x \partial y} \right)^2 < 0$$

at the critical point, then we know the critical point is a saddle point. We summarize all of this in the following test:

Test for Extrema for Functions of Two Variables:

Given a function of two variables $f(x, y)$ the critical points of this function are the points (x, y) that satisfy

$$\frac{\partial f}{\partial x} = 0, \quad \text{and} \quad \frac{\partial f}{\partial y} = 0.$$

Suppose that at the critical point(s), we have

$$D = \frac{\partial^2 f}{\partial x^2} \times \frac{\partial^2 f}{\partial y^2} - \left(\frac{\partial^2 f}{\partial x \partial y}\right)^2 > 0,$$

then if

$$\frac{\partial^2 f}{\partial x^2} > 0$$

the critical point is a minimum and if

$$\frac{\partial^2 f}{\partial x^2} < 0$$

the critical point is a maximum. If $D < 0$ at the critical point, it is a saddle point. If $D = 0$ at the critical point then this test is inconclusive.

Example 8.30

Find the extrema of the function $f(x, y) = 3x^2 - 4y^2 - 6x + 1$.

Solution: First we find $\frac{\partial f}{\partial x} = 6x - 6$ and $\frac{\partial f}{\partial y} = -8y$. Then we solve $\frac{\partial f}{\partial x} = 0$ and $\frac{\partial f}{\partial y} = 0$, which gives us $x = 1$ and $y = 0$. Therefore, there is a critical point at $(1, 0)$. Next, we find

$$\frac{\partial^2 f}{\partial x^2} = 6, \qquad \frac{\partial^2 f}{\partial y^2} = -8, \qquad \frac{\partial^2 f}{\partial x \partial y} = 0.$$

Substitution into our condition gives us

$$D = \frac{\partial^2 f}{\partial x^2} \times \frac{\partial^2 f}{\partial y^2} - \left(\frac{\partial^2 f}{\partial x \partial y}\right)^2 = 6(-8) - 0^2 = -48 < 0.$$

Hence, the critical point is not a local minimum or maximum, it is a saddle point.

Example 8.31

Find the extrema of the function $f(x, y) = 2x^2 + 3y^2$.

Solution: First we find $\frac{\partial f}{\partial x} = 4x$ and $\frac{\partial f}{\partial y} = 6y$. Then we solve $\frac{\partial f}{\partial x} = 0$ and $\frac{\partial f}{\partial y} = 0$, which gives us $x = 0$ and $y = 0$. Therefore, there is a critical point at $(0,0)$. Next, we find

$$\frac{\partial^2 f}{\partial x^2} = 4, \qquad \frac{\partial^2 f}{\partial y^2} = 6, \qquad \frac{\partial^2 f}{\partial x \partial y} = 0.$$

Substitution into our condition gives us

$$D = \frac{\partial^2 f}{\partial x^2} \times \frac{\partial^2 f}{\partial y^2} - \left(\frac{\partial^2 f}{\partial x \partial y}\right)^2 = 4(6) - 0^2 = 24 > 0.$$

Since $\frac{\partial^2 f}{\partial x^2} = 4 > 0$, the function f has a minimum at the point $(0,0)$.

Example 8.32

Find the extrema of the function $f(x,y) = 3x^2 + xy - y^2 - x^3$.

Solution: First we find $\frac{\partial f}{\partial x} = 6x + y - 3x^2$ and $\frac{\partial f}{\partial y} = x - 2y$. Then we solve $\frac{\partial f}{\partial x} = 0$ and $\frac{\partial f}{\partial y} = 0$. This gives us the following system of two equations with two unknowns:

$$6x + y - 3x^2 = 0$$
$$x - 2y = 0$$

From the second equation we obtain $y = \frac{x}{2}$, which we then substitute into the first equation to obtain

$$6x + \frac{x}{2} - 3x^2 = 0 \implies 12x + x - 6x^2 = 0 \implies x(13 - 6x) = 0.$$

Therefore $x = 0$ or $x = \frac{13}{6}$. Using this we find that when $x = 0$, $y = 0$, and when $x = \frac{13}{6}$, $y = \frac{13}{12}$. At the first critical point $(0,0)$ we get:

$$\frac{\partial^2 f}{\partial x^2} = 6, \qquad \frac{\partial^2 f}{\partial y^2} = -2, \qquad \frac{\partial^2 f}{\partial x \partial y} = 1.$$

Therefore, $D = \frac{\partial^2 f}{\partial x^2} \times \frac{\partial^2 f}{\partial y^2} - \left(\frac{\partial^2 f}{\partial x \partial y}\right)^2 = -13 < 0$. There is no minimum nor maximum at $(0,0)$.

At the second critical point $\left(\frac{13}{6}, \frac{13}{12}\right)$ we get:

$$\frac{\partial^2 f}{\partial x^2} = -7, \qquad \frac{\partial^2 f}{\partial y^2} = -2, \qquad \frac{\partial^2 f}{\partial x \partial y} = 1.$$

Therefore, $D = \frac{\partial^2 f}{\partial x^2} \times \frac{\partial^2 f}{\partial y^2} - \left(\frac{\partial^2 f}{\partial x \partial y}\right)^2 = 13 > 0$. Therefore this critical point is either a maximum or a minimum. Since $\frac{\partial^2 f}{\partial x^2} = -7 < 0$, we have that f has a maximum at the point $\left(\frac{13}{6}, \frac{13}{12}\right)$.

8.5 PROBLEMS

Question 8.1 *Evaluate the below functions at the given points:*

(a) $f(x,y) = 2x - 3y$ at $x=2$, $y=-1$

(b) $f(x,y) = 3x^2 y^4$ at $x=4$, $y=2$

(c) $f(x,y) = 3^x + 4^y$ at $x=1$, $y=2$

(d) $f(x,y) = 3xy - \frac{10}{y}$ at $x=2$, $y=5$

(e) $f(x,y) = x - e^y$ at $x=-1$, $y=0$

(f) $f(x,y) = e^{xy} + 7$ at $x = -1$, $y = -3$

(g) $f(x,y) = \ln(2xy)$ at $x=2$, $y=3$

(h) $f(x,y) = 5x^2 + 3y - \ln y$ at $x=0$, $y=1$

Question 8.2 *Find both first-order partial derivatives of the following functions:*

(a) $f(x,y) = 5x^2 - 15xy + 2y^3$

(b) $f(x,y) = 2x + 9x^2 y^2 - 5y^2$

(c) $f(x,y) = 6x + 9x^2 y^2 - 8y^2$

(d) $f(x,y) = 5x^2 - 6x^2 y^2 + 7y$

(e) $f(x,y) = -3x^2 + 4x^2 y^2 - 9y$

(f) $f(x,y) = 8x^3 y - 7y^2 + 2x$

(g) $f(x,y) = 5xy^3 + 6x^2 + 3y$

(h) $f(x,y) = 4x^4 - 10x^3 y^3 + 7y^4$

Question 8.3 *Find both first-order partial derivatives of the following functions:*

(a) $f(x,y) = (3x^2 + y^4)^4$

(b) $f(x,y) = (-5x^3 - 4y^2)^7$

(c) $f(x,y) = (x^2 y + y^2 x)^3$

(d) $f(x,y) = (3x^2 + 5x^3 y^3 - 2y^2)^5$

(e) $f(x,y) = (4x^3 + 3y^4)^{-2}$

(f) $f(x,y) = (-3x^2 y + 2xy^3)^{-4}$

(g) $f(x,y) = \frac{1}{(10x^5 - 4y^7)^2}$

(h) $f(x,y) = \frac{1}{(x^2 - y^2)^3}$

Question 8.4 *Find both first-order partial derivatives of the following functions:*

(a) $f(x,y) = e^{xy}$

(b) $f(x,y) = e^{x-y}$

(c) $f(x,y) = 5e^{x^2 y^3}$

(d) $f(x,y) = 3e^{x^3 + y^2}$

(e) $f(x,y) = \ln(2xy)$

(f) $f(x,y) = \ln(3x^2 + 2y^3)$

(g) $f(x,y) = 7\ln(x-y)$

(h) $f(x,y) = 10\ln(2x^3 y^3)$

Question 8.5 *Find both first-order partial derivatives of the following functions:*

(a) $f(x,y) = (x-y)(x-y)$

(b) $f(x,y) = (2x+y)(3x^2 - 2y^3)$

(c) $f(x,y) = (5x^2)(4x + 5y - 10)$

(d) $f(x,y) = (3x^2 y^2 + 5)(4x^3 - 2x^2)$

(e) $f(x,y) = \frac{y}{x}$

(f) $f(x,y) = \frac{y-x}{x-y}$

(g) $f(x,y) = \frac{2x^2}{5x+2y}$

(h) $f(x,y) = \frac{2x-3y}{4x^2 y^3}$

Question 8.6 *Find all second-order partial derivatives of the following functions:*

(a) $f(x,y) = 5x^3y^5$

(b) $f(x,y) = 3x^7 - 4y^5$

(c) $f(x,y) = (2x^3 - 5y)^2$

(d) $f(x,y) = (x^2 + y^2)^3$

(e) $f(x,y) = (3x^2 - y^3)^{-1}$

(f) $f(x,y) = \frac{1}{x+3y^2}$

(g) $f(x,y) = e^{x^2y^2}$

(h) $f(x,y) = e^{x^2+y^2}$

(i) $f(x,y) = \ln(2x^3y^3)$

(j) $f(x,y) = \ln(2x^3 + y^3)$

Question 8.7 *Consider* $z = (2x^3 - y^2)^2$.

1. *Find the differential of z.*
2. *Use the differentials to find the approximate change in z if x decreases by 3% and y increases by 2%.*

Question 8.8 *Consider* $z = 3e^{x^3+y^2}$.

1. *Find the differential of z.*
2. *Use the differentials to find the approximate change in z if x increases by 2% and y deceases by 4%.*

Question 8.9 *If the production function of a good is given by: $z = xy - 2x^3y^2 + 3x - 1$, find the marginal productivity of x and y.*

Question 8.10 *If the production function of a good is given by: $z = x^x - 2y^2 + \frac{3}{x} + 2$, find the marginal productivity of x and y. (Hint: Write $x^x = e^{\ln(x^x)} = e^{x\ln(x)}$.)*

Question 8.11 *Suppose a company makes two different types of widgets, type X and type Y. Each week the company sells x type X widgets and y type Y widgets. The company has the following cost and revenue functions:*

$$C(x,y) = 7x + 9y + 18,000,$$
$$R(x,y) = 50x + 70y + 0.07xy - 0.04x^2 - 0.04y^2.$$

Find the profit function for the company and then find $P_x(800, 1000)$ and $P_y(800, 1000)$ and interpret their meaning.

Question 8.12 *Suppose a company makes two products, X and Y. The price demand functions of X and Y are given by $p_1 = 250 - 2x + 2y$ and $p_2 = 300 + 3x - y$ respectively. The cost function is $C(x,y) = 400 + 40x + 200y$. Find $P_x(300, 200)$ and $P_y(300, 200)$ and interpret what these values mean.*

CHAPTER 9

Special Topics

In this chapter, we will take a look at a few special mathematical and economic topics. On the mathematical side of things, we will look at implicit differentiation, L'Hôpital's rule, and the method of Lagrange multipliers. These are all mathematical techniques that are extremely useful in economics. On the economics side of things we will look at the concept of elasticity of demand and the important Cobb-Douglas production function. Both of these are important in economic theory.

9.1   IMPLICIT DIFFERENTIATION

So far we have always worked with functions having the form $y = f(x)$. Here, the dependent variable y is given as an explicit formula in terms of the independent variable, usually x. In other words, the variable y is equal to some algebraic expression, which is in terms of the variable x. This type of function is called an **explicit function**. However, it is possible that both variables y and x appear on the same side of an equation together. Functions like this are called **implicit functions**. However, we need to be careful. Not all expressions that look like implicit functions are actually functions.

For example, $y = x^2 - 2x + 1$ is an explicit function while $x^2 - 4y + 5 = 0$ is an implicit equation. We do not necessarily know if this equation is actually a function. Sometimes, it is possible to solve an implicit equation to obtain an explicit function. Then we would know that the implicit equation was, in fact, an implicit function. In this case, it is relatively easy to show that $x^2 - 4y + 5 = 0$ is indeed an implicit function since we can write $x^2 - 4y + 5 = 0$ in the explicit form as

$$x^2 - 4y + 5 = 0 \implies 4y = x^2 + 5$$
$$\implies y = \frac{x^2 + 5}{4}.$$

Each input value x gives only one output value y, which means that y is indeed a function of x. Thus we could write $y = f(x) = \frac{x^2+5}{4}$.

DOI: 10.1201/9781003480235-9

Now let us consider a different implicit equation, $x^2 + y^2 - 4 = 0$. If we tried to solve for y, we would get

$$x^2 + y^2 - 4 = 0 \implies y^2 = 4 - x^2$$
$$\implies y = \pm\sqrt{4 - x^2}.$$

Here, we could find an input value of x that would give more than one output value y; for example, if we choose $x = 1$, we would have $y = \sqrt{3}$ and $y = -\sqrt{3}$. This violates the definition of a function, hence y is not a function of x.

Indeed, if we had chosen to solve our original equation $x^2 - 4y + 5 = 0$ for x, we would have obtained

$$x^2 - 4y + 5 = 0 \implies x^2 = 4y - 5$$
$$\implies x = \pm\sqrt{4y - 5}.$$

We can see clearly that x is not a function of y.

Finding the derivative of a function that is defined implicitly is called **implicit differentiation**. This is done using the chain rule and viewing the variable y as being an implicit function of x. We think of y as somehow depending on x, that is, $y = y(x)$. Suppose we wanted to find the derivative $\frac{d(y^2)}{dx}$. Using the chain rule, we would first find the derivative of y^2 with respect to the variable y, which is $2y$. Then we would find the derivative of $y(x)$ with respect to the variable x. Since we do not actually know the exact representation of y in terms of x the best we can do is write $\frac{dy}{dx}$. Combining this with the chain rule we get the derivative of y^2 with respect to x to be

$$\frac{d(y^2)}{dx} = \frac{d(y^2)}{dy} \cdot \frac{dy}{dx} = 2y\frac{dy}{dx},$$

which we might also write as

$$\left(y^2\right)' = 2yy'.$$

Now let us try to implicitly differentiate the implicit function $x^2 - 4y + 5 = 0$ with respect to x. (Since we know that y is an implicit function of x). We have

$$\frac{d}{dx}\left(x^2 - 4y + 5 = 0\right) \implies \frac{d}{dx}\left(x^2 - 4y + 5\right) = \frac{d}{dx}\left(0\right)$$
$$\implies \frac{d(x^2)}{dx} - \frac{d(4y)}{dx} + \frac{d(5)}{dx} = \frac{d(0)}{dx}.$$

Since x^2 is already written in terms of x, we have $\frac{d(x^2)}{dx} = 2x$ and since both 5 and 0 are constants, we have $\frac{d(5)}{dx} = \frac{d(0)}{dx} = 0$. We also have

$$\frac{d(4y)}{dx} = \frac{d(4y)}{dy} \cdot \frac{dy}{dx} = 4 \cdot \frac{dy}{dx} = 4y'.$$

Putting all of this together, we have

$$\frac{d}{dx}\left(x^2 - 4y + 5 = 0\right) \implies 2x - 4y' = 0,$$

which we can algebraically solve for y' to get

$$y' = \frac{x}{2}.$$

In this case, we can double check our answer. Since the implicitly defined function $x^2 - 4y + 5 = 0$ could be written as an explicit function $y = \frac{x^2+5}{4}$ would could just take the derivative of the function using already known techniques. And indeed, when we do we find that $y' = \frac{x}{2}$, which checks out. Implicit differentiation is most useful when we can not write an implicitly defined function as an explicitly defined function.

Example 9.1

Given $x^2y - 2x - 7 = 0$, find y' using implicit differentiation.

Solution: By differentiating implicitly both sides of the equation we obtain:

$$\frac{d}{dx}\left(x^2y - 2x - 5 = 0\right) \implies \frac{d}{dx}\left(x^2y - 2x - 5\right) = \frac{d}{dx}\left(0\right)$$

$$\implies \frac{d(x^2y)}{dx} - \frac{d(2x)}{dx} - \frac{d(5)}{dx} = \frac{d(0)}{dx}.$$

It is clear that $\frac{d(2x)}{dx} = 2$ and $\frac{d(5)}{dx} = \frac{d(0)}{dx} = 0$. It is the first term we need to focus on. Since we are assuming that y is implicitly defined in terms of x we could write $y = y(x)$. If we did that then the term x^2y would actually be the product of two functions of x, x^2 and $y(x)$. This would necessitate using the product rule;

$$\frac{d(x^2y)}{dx} = \frac{d(x^2)}{dx} \cdot y + x^2 \cdot \frac{dy}{dx} = 2xy + x^2y'.$$

Putting all this together, we have

$$\frac{d}{dx}\left(x^2y - 2x - 5 = 0\right) \implies 2xy + x^2y' - 2 = 0$$

$$\implies x^2y' = 2 - 2xy$$

$$\implies y' = \frac{2 - 2xy}{x^2}.$$

Notice, in this particular case we could actually have solved this implicit equation as $y = \frac{2x+7}{x^2}$ and taken the derivative using the quotient rule.

Example 9.2

Given $x^2 + 2x - y^3 = 4$, find y' and the slope of the tangent line to the graph at $x = 1$.

Solution: Differentiating implicitly both sides of the equation $x^2 + 2x - y^3 = 4$ gives us

$$\frac{d}{dx}\left(x^2 + 2x - y^3 = 4\right) \implies \frac{d}{dx}\left(x^2 + 2x - y^3\right) = \frac{d}{dx}(4)$$

$$\implies \frac{d(x^2)}{dx} + \frac{d(2x)}{dx} - \frac{d(y^3)}{dx} = \frac{d(4)}{dx}$$

$$\implies 2x + 2 - 3y^2 y' = 0$$

$$\implies y' = \frac{2x + 2}{3y^2}.$$

The slope of the tangent line to the graph at $x = 1$ is the value of the derivative y' at $x = 1$. We first need to find the y value for $x = 1$ by substituting in the equation $x^2 + 2x - y^3 = 4$

$$\text{at } x = 1 \implies (1)^2 + 2(1) - y^3 = 4$$

$$\implies 3 - y^3 = 4$$

$$\implies y^3 = -1$$

$$\implies y = -1.$$

We can then use $x = 1$ and $y = -1$ to find the slope of the tangent line $y' = \frac{4}{3}$. So the slope at the point $(1, -1)$ is $y' = \frac{4}{3}$.

Example 9.3

Find y'' for $x^3 + 4y^2 = 5$.

Solution: By using implicit differentiation on $x^3 + 4y^2 = 5$ we find $3x^2 + 8yy' = 0$. Now we use implicit differentiation for the second time:

$$\frac{d}{dx}\left(3x^2 + 8yy' = 0\right) \implies 6x + 8(y'y' + yy'') = 0$$

$$\implies 8yy'' = -8y'y' - 6x$$

$$\implies y'' = \frac{-8(y')^2 - 6x}{8y}.$$

Notice how we needed to use the product rule on the yy' term.

9.2   L'HÔPITAL'S RULE

Limits play a very important role in derivatives and graphing functions as well. In Chapter 2, we saw how it was possible to find limits of functions. Using the derivative, we will now introduce a more general procedure for determining the limit of functions of indeterminate form $\frac{0}{0}$. This procedure is based on L'Hôpital's Rule.

Example 9.4

Find the limit of the following function when x approaches 3:

$$h(x) = \frac{x^2 - 9}{x - 3}.$$

Solution: If we try just substituting 3 directly in the function we obtain $h(3) = \frac{0}{0}$, which is an indeterminate form. Being a little more mathematically precise, if we look at both the numerator and denominator of $h(x)$, $f(x) = x^2 - 9$ and $g(x) = x - 3$, we notice that

$$\lim_{x \to 3} h(x) = \frac{\lim_{x \to 3} f(x)}{\lim_{x \to 3} g(x)},$$

Then both $\lim_{x \to 3} f(x) = \lim_{x \to 3}(x^2 - 9) = 0$ and $\lim_{x \to 3} g(x) = \lim_{x \to 3}(x - 3) = 0$. Therefore $\lim_{x \to 3} h(x) = \frac{0}{0}$, which is an indeterminate form. In this particular case we realize that we can simplify $h(x)$ to obtain the following:

$$h(x) = \frac{x^2 - 9}{x - 3} = \frac{\cancel{(x - 3)}(x + 3)}{\cancel{x - 3}} = (x + 3).$$

This eliminates the issue of 0 in the denominator and allows us to find the limit,

$$\lim_{x \to 0} h(x) = \lim_{x \to 0}(x + 3) = 6.$$

Unfortunately, this trick of simplifying the quotient by factoring the numerator and canceling with the denominator does not work in general. For this reason we introduce L'Hôpital's Rule.

L'Hôpital's Rule for limits of indeterminate form $\frac{0}{0}$:

Suppose a is real number, f and g are two differentiable functions, and $g'(a) \neq 0$. If $\lim_{x \to a} f(x) = \lim_{x \to a} g(x) = 0$, then

$$\lim_{x \to a} \frac{f(x)}{g(x)} = \lim_{x \to a} \frac{f'(x)}{g'(x)}.$$

If $\lim_{x \to a} \frac{f'(x)}{g'(x)}$ is also an indeterminate form $\frac{0}{0}$, we can apply L'Hôpital's Rule a second time

$$\lim_{x \to a} \frac{f'(x)}{g'(x)} = \lim_{x \to a} \frac{f''(x)}{g''(x)}.$$

It is important to note that L'Hôpital's Rule also applies to limits of the indeterminate form $\frac{\pm \infty}{\pm \infty}$

Example 9.5

Find the limit of the following function when x approaches 5:

$$f(x) = \frac{\sqrt{x+4} - 3}{\sqrt{x-5}}$$

Solution: Since $\lim_{x \to 5} f(x) = \frac{0}{0}$, we apply L'Hôpital's Rule as follows:

$$
\begin{aligned}
\lim_{x \to 5} f(x) &= \lim_{x \to 5} \frac{\sqrt{x+4} - 3}{\sqrt{x-5}} \\
&= \lim_{x \to 5} \frac{(x+4)^{\frac{1}{2}} - 3}{(x-5)^{\frac{1}{2}}} \\
&= \lim_{x \to 5} \frac{\left((x+4)^{\frac{1}{2}} - 3\right)'}{\left((x-5)^{\frac{1}{2}}\right)'} \\
&= \lim_{x \to 5} \frac{\frac{1}{2}(x+4)^{-\frac{1}{2}}}{\frac{1}{2}(x-5)^{-\frac{1}{2}}} \\
&= \lim_{x \to 5} \frac{\sqrt{x-5}}{\sqrt{x+4}} \\
&= \frac{0}{3} \\
&= 0.
\end{aligned}
$$

Example 9.6

Find the limit of the following function when x approaches 1:

$$f(x) = \frac{x^2 - \frac{1}{x}}{x - 1}.$$

Solution: Since $\lim_{x\to 1} f(x) = \frac{0}{0}$, we apply L'Hôpital's Rule as follows:

$$\lim_{x\to 1} f(x) = \lim_{x\to 1} \frac{x^2 - \frac{1}{x}}{x - 1}$$

$$= \lim_{x\to 1} \frac{\left(x^2 - \frac{1}{x}\right)'}{(x - 1)'}$$

$$= \lim_{x\to 1} \frac{2x + \frac{1}{x^2}}{1}$$

$$= \lim_{x\to 1} \left(2x + \frac{1}{x^2}\right)$$

$$= 3.$$

Example 9.7

Find the limit of the following function when x approaches ∞:

$$f(x) = \frac{x^3 - 3}{x^2 - 5x}$$

Solution: Since $\lim_{x\to\infty} f(x) = \frac{\infty}{\infty}$, we apply L'Hôpital's Rule as follows:

$$\lim_{x\to\infty} f(x) = \lim_{x\to\infty} \frac{x^3 - 3}{x^2 - 5x}$$

$$= \lim_{x\to\infty} \frac{(x^3 - 3)'}{(x^2 - 5x)'}$$

$$= \lim_{x\to\infty} \frac{3x^2}{2x - 5}$$

$$= \frac{\infty}{\infty}.$$

Therefore we apply L'Hôpital's Rule again:

$$\lim_{x\to\infty} f(x) = \lim_{x\to\infty} \frac{3x^2}{2x - 5}$$

$$= \lim_{x\to\infty} \frac{(3x^2)'}{(2x - 5)'}$$

$$= \lim_{x\to\infty} \frac{6x}{2}$$

$$= \lim_{x\to\infty} 3x$$

$$= \infty.$$

9.3   ELASTICITY OF DEMAND

In economics, we are often interested in how price changes affect demand and revenue. To study this relationship between price, demand, and revenue, economists use a concept called Elasticity of Demand. Let us consider the revenue function $R(p) = p \cdot x$ where p is given by the price-demand function $p = D(x)$. We have

$$R(p) = p \cdot x = D(x) \cdot x$$

A change in price results in a change in demand, which is measured by the slope of the demand function $D'(x)$. The price-demand function $p = D(x)$ was discussed in Chapter 5. But notice, we could also have written the price-demand function as quantity in terms of price; $x = D(p)$.

Example 9.8

We are given the price demand function $p = 1000 - 0.1x$. Write this price-demand function as quantity in terms of price.

Solution: All we need to do is solve for x,

$$p = 1000 - 0.1x \implies 0.1x = 1000 - p$$

$$\implies x = \frac{1000 - p}{0.1}$$

$$\implies x = 10000 - 10p.$$

Thus, we need to pay close attention to how the price-demand function is written.

For the following, we will assume that the price-demand function is given in the form $x = D(p)$. Notice that this is different than the way the price-demand function was given in Chapter 5. The **price elasticity of demand**, $e(p)$, is defined as the percentage change in demand for each one percent change in price p. This is obtained by dividing the percent change in demand, given by $\frac{\Delta x}{x}$, with the percent change in price, given by $\frac{\Delta p}{p}$;

$$\epsilon(p) = \frac{\frac{\Delta x}{x}}{\frac{\Delta p}{p}} = \frac{\Delta x}{x} \cdot \frac{p}{\Delta p} = \frac{\Delta x}{\Delta p} \cdot \frac{p}{x}$$

As we saw in Chapter 2, $\frac{\Delta x}{\Delta p}$ is, in the limit, the marginal demand $D'(p)$, therefore we can rewrite this as

$$\epsilon(p) = D'(p)\frac{p}{x}$$

and since $x = D(p)$, we have

$$\epsilon(p) = \frac{D'(p)}{D(p)}p.$$

Sometimes the price elasticity of demand is simply called the elasticity of demand. By looking at the elasticity formula, we notice that $D'(p)$ is always negative as demand is a decreasing function. Since p and $x = D(p)$ are always positive, we obtain that elasticity $\epsilon(p)$ is always negative. But what exactly does elasticity of demand measure? In essence, it measures the percent change in demand given a one percent increase in price. Since elasticity of demand is negative, this means that a one percent increase in price will result in a decrease in demand, as would be expected from economic theory.

A word of caution; sometimes you will see the elasticity of the demand function defined as

$$\epsilon(p) = -\frac{D'(p)}{D(p)}p.$$

This extra negative sign is often used by economists because in general one would expect an increase in price to result in a decrease in demand. Thus, the value $\epsilon(p)$ would give the expected decrease in demand associated with a 1% increase in price as a positive number. When defined like this $\epsilon(p)$ is always positive. When you encounter elasticity of demand, it is always important to check which definition is being used, with or without the negative sign.

The Elasticity of Demand:

The elasticity of demand of a good with demand function $x = D(p)$ is the percent change in demand for each one percent change in price and is given by the following formula

$$\epsilon(p) = \frac{D'(p)}{D(p)}p.$$

- A good is said to be **inelastic** if its price elasticity, $\epsilon(p)$ is between -1 and 0. In this case, a price change has a weak effect on demand.
- A good is said to be **elastic** if its price elasticity, $\epsilon(p)$ is below -1. In this case, a price change has a strong effect on demand.
- A good is said to be **unit elastic** if its price elasticity, $\epsilon(p)$ is -1. In this case, the percent change in demand is the same as the percent change in price.

Note, if you were using the second formula for elasticity of demand, with the negative sign, all the -1s in these definitions would be 1.

Example 9.9

Consider the price-demand function $x = 3 - 2p$. Find the elasticity of demand, $\epsilon(p)$, and interpret $\epsilon(1)$.

Solution: As $x = D(p) = 3 - 2p \implies D'(p) = -2$ so

$$\epsilon(p) = \frac{D'(p)}{D(p)}p = \frac{-2}{3 - 2p}p = \frac{-2p}{3 - 2p}.$$

Thus, at the price of $1, $\epsilon(1) = -2$, which is below -1, so demand is elastic. At the price of $1, a 1% increase in price, will cause demand to decrease by 2%.

Now consider the revenue function defined in terms of price p instead of quantity x,

$$R(p) = p \cdot x = p \cdot D(p).$$

If we differentiate with respect to p, using the product rule, and then divide and multiply the second term with $D(p)$, we obtain

$$R'(p) = D(p) + D'(p) \cdot p$$
$$= D(p) + \frac{D'(p) \cdot p}{D(p)} D(p)$$
$$= D(p) + \epsilon(p) D(p)$$
$$= D(p)[1 + \epsilon(p)].$$

As $D(p)$ is always positive, $R'(p)$ and $1 + \epsilon(p)$ have same sign.

- When a good is inelastic, i.e., its price elasticity $\epsilon(p)$ is between -1 and 0, this implies that $R'(p)$ is positive and therefore an increase in price will result in an increase in the revenue.
- When a good is elastic, i.e., its price elasticity, $\epsilon(p)$ is below -1, this implies that $R'(p)$ is negative and therefore an increase in price will result in a decrease in the revenue.
- A good is said to be **unit elastic** if its price elasticity, $\epsilon(p)$ is -1, this implies that $R'(p) = 0$, and therefore, the revenue is unchanged in this case.

Example 9.10

Find the elasticity of demand of a commodity with demand function $x = 500 - 10p$. Suppose the commodity has a price of $10. What happens to revenue if the price increases 1%?

Solution: As $x = D(p) = 500 - 10p \implies D'(p) = -10$.

$$\epsilon(p) = \frac{D'(p)}{D(p)} p$$

$$= \frac{-10p}{500 - 10p}$$

at the price of $10, $\epsilon(10) = -0.25$, which is between -1 and 0, so demand is inelastic. As $1 + \epsilon(10) > 0$, it implies that $R'(p)$ is also positive. At the price of $10, an increase in price will increase revenue.

Example 9.11

The demand function for a certain product is given by

$$D(p) = \frac{1200}{p},$$

where $D(p)$ is the quantity demanded, and p is the price per unit. Calculate the price elasticity of demand when the price is $30.

Solution: Price elasticity of demand is given by

$$E(p) = p \cdot \frac{D'(p)}{D(p)}.$$

First, we find the derivative of the demand function with respect to p.

$$D'(p) = \frac{d}{dp} \left(\frac{1200}{p} \right)$$

$$= -\frac{1200}{p^2}.$$

Then, we calculate the elasticity at $p = 30$.

$$E(30) = 30 \times \frac{-1200/30^2}{1200/30}$$

$$= 30 \times \frac{-1200}{900 \times 40}$$

$$= -1.$$

A price elasticity of demand of -1 indicates unitary elasticity; the quantity demanded changes by the same percentage as the price.

9.4   COBB-DOUGLAS PRODUCTION FUNCTION

The Cobb-Douglas production function is a fundamental concept in both macroeconomics and production economics, serving as a crucial tool in understanding the relationship between inputs and output in production processes. The Cobb-Douglas production function is primarily used to model the output of a production process by relating the quantity of output to the quantities of capital and labor inputs. In economic theory, this function is used to explore how firms can maximize production under certain constraints or minimize costs for a given level of production.

From the perspective of producer theory, a firm aims to optimize its production function while operating under certain constraints. Specifically, by applying a cost function constraint, the firm can effectively minimize production costs while adhering to its budgetary limitations. This strategy allows the firm to maximize its level of production according to the cost it has.

The Cobb-Douglas production function is mathematically represented as:

$$Y = F(K, L) = A \cdot K^\alpha \cdot L^\beta$$

where:

- Y is the total production (the real value of all goods produced in a year),
- A is total factor productivity,
- K is the amount of capital invested,
- L is the amount of labor used,
- $0 < \alpha < 1$ and $0 < \beta < 1$ represent the output elasticities of capital and labor, respectively, which measure the percentage increase in output resulting from a one percent increase in capital or labor.

It is worth noting again that α and β are the output elasticities of capital and labor, respectively, it means if we compute the elasticity of demand for capital we will find α and elasticity of demand for labor, we will find β.

The Cobb-Douglas production function helps economists analyze the elasticity of output relative to labor and capital, and understand the concept of returns to scale in production. In practice, the Cobb-Douglas model is often extended or modified to include additional factors like human capital, technology, and natural resources. These enhancements acknowledge the complex nature of production in modern economies and the role of innovation and resource management in growth.

Example 9.12

Suppose a company's production function is given by the Cobb-Douglas form

$$Y = 10 \cdot K^{0.7} \cdot L^{0.3}$$

where Y is the total output, K is the capital input, L is the labor input, and the coefficient 10 represents a productivity factor that scales the output given the inputs. The exponents $\alpha = 0.7$ and $\beta = 0.3$ on K and L are the output elasticities of capital and labor, respectively. Calculate the output if the company employs 100 units of labor and 100 units of capital. What is the output if the company doubles labor and capital to 200 units each?

Solution: If the company employs 100 units of labor and 100 units of capital, we can substitute these values into the production function to calculate output we get

$$
\begin{aligned}
Y &= 10 \cdot 100^{0.7} \cdot 100^{0.3} \\
&= 10 \cdot (10^2)^{0.7} \cdot (10^2)^{0.3} \\
&= 10 \cdot 10^{1.4} \cdot 10^{0.6} \\
&= 10 \cdot 10^2 \\
&= 1000.
\end{aligned}
$$

Next, if we double labor to 200 units and capital to 200 units, we have

$$
\begin{aligned}
Y &= 10 \cdot 200^{0.7} \cdot 200^{0.3} \\
&= 10 \cdot (2 \cdot 10^2)^{0.7} (2 \cdot 10^2)^{0.3} \\
&= 10 \cdot 2^{0.7} \cdot 10^{1.4} \cdot 2^{0.3} \cdot 10^{0.6} \\
&= 10 \cdot 2^{0.7+0.3} \cdot 10^2 \\
&= 2000.
\end{aligned}
$$

So we find out that doubling both inputs (capital and labor) leads to a doubling of output, which is a property known as **constant returns to scale**. It is important to highlight that this property occurs because the sum of the output elasticities is equal to 1, i.e., $\alpha + \beta = 1$.

Cobb-Douglas Production Function $Y = 10K^{0.7}L^{0.3}$

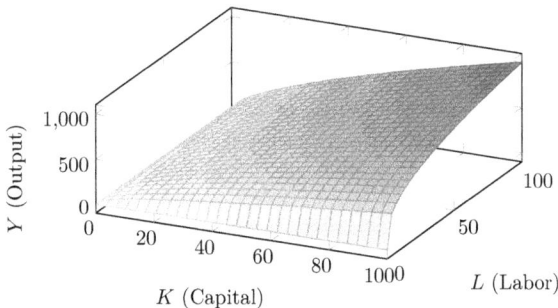

Example 9.13

Let us change the above example slightly by considering a different value for β. Suppose a company's production function is given by

$$Y = 10 \cdot K^{0.7} \cdot L^{0.2}.$$

Determine how the production scales with $\beta = 0.2$ as opposed to $\beta = 0.3$ as in the last example.

Solution: If the company employs 100 units of labor and 100 units of capital, the output can be calculated as

$$Y = 10 \cdot (10^2)^{0.7} \cdot (10^2)^{0.2} = 10 \cdot 10^{1.8} \approx 631.$$

Compare that to the answer we obtained in the last example for the same number of units of capital and labor, $Y = 1000$. The reduced output indicates how the production scale adjusts with reduced elasticity of labor compared to the original scenario. In this case, we notice that the sum of the exponents ($\alpha + \beta = 0.9$) is less than 1, which reflects **decreasing returns to scale**.

Example 9.14

Let us again change our example by changing both α and β values. Suppose a company's production function is given by

$$Y = 10 \cdot K^{0.6} \cdot L^{0.5}.$$

Determine how the production scales in comparison to the previous examples.

Solution: If the company employs 100 units of labor and 100 units of capital, the output can be calculated as

$$Y = 10 \cdot (10^2)^{0.6} \cdot (10^2)^{0.5} = 10 \cdot 10^{2.2} \approx 1585$$

In this scenario, the output indicates an increased sensitivity to changes in both capital and labor inputs compared to previous examples. The sum of the exponents, $\alpha + \beta = 1.1$, exceeds 1, indicating **increasing returns to scale**.

In general, for any scale λ, the Cobb-Douglas production function is homogeneous of degree $(\alpha + \beta)$. In other words,

$$F(\lambda K, \lambda L) = A \cdot (\lambda K)^{\alpha} \cdot (\lambda L)^{\beta} = A\lambda^{\alpha}K^{\alpha}\lambda^{\beta}L^{\beta} = \lambda^{\alpha+\beta}F(K, L).$$

Furthermore,

> If $\alpha + \beta < 1$, we have decreasing returns to scale.
> If $\alpha + \beta = 1$, we have constant returns to scale.
> If $\alpha + \beta > 1$, we have increasing returns to scale.

In production theory, we distinguish between short-term and long-term in terms of production behavior. In the short-term we usually think of one of the factors of production remaining constant and the other factor changing. In general, for the short-term we usually think of capital (K) as the factor that remains unchanged. This is because capital is usually required when opening closing product lines, which is something that is not straightforward and usually takes time to do. Also, in the short-term, we generally think of the labor factor (L) as much more likely to vary since hiring or firing employees is relatively easy to do.

On the other hand, in the long-term, both production factors capital and labor can vary, and here we can apply the optimization tools we have learned in previous chapters. We can look for an optimum solution where production firms try to maximize their production under some constraints, such as average productivity, capital marginal productivity, labor marginal productivity, law of diminishing returns, or returns to scale that was already seen in the examples above.

The average productivity for each factor of production, labor or capital, is the quantity of production produced per unit of that factor:

$$AP_K = \frac{Y}{K} \quad \text{and} \quad AP_L = \frac{Y}{L}.$$

Marginal productivity is simply the additional production induced by the increase of one unit of labor or capital, as seen in previous chapters. Thus, marginal productivity is simply the first derivative of the production function. Therefore, the **marginal productivity of capital** is given by

$$MP_K = \frac{\partial Y}{\partial K} = \frac{\partial (A \cdot K^\alpha \cdot L^\beta)}{\partial K} = A\alpha K^{\alpha-1} L^\beta$$

We can simplify MP_K as follows:

$$MP_K = \alpha \frac{A K^{\alpha-1} L^\beta K}{K} = \alpha \frac{A K^\alpha L^\beta}{K} . = \alpha \frac{Y}{K} = \alpha \cdot AP_K.$$

It is important to note that since capital K, production Y, and α are all positive, we find out that the marginal product of capital is always positive; that is, increasing capital leads to an increase in output. In a similar way, the **marginal productivity of labor** is given by

$$MP_L = \frac{\partial Y}{\partial L} = \frac{\partial (A \cdot K^\alpha \cdot L^\beta)}{\partial L} = A\beta K^\alpha L^{\beta-1},$$

which can be similarly simplified as

$$MP_L = \beta\frac{AK^\alpha L^{\beta-1}L}{L} = \beta\frac{AK^\alpha L^\beta}{L} = \beta\frac{Y}{L} = \beta \cdot AP_L.$$

In economics, particularly when dealing with production functions like the Cobb-Douglas model, we use what is called the **Marginal Rate of Technical Substitution** (MRTS) to measure how much of one input, labor for example, needs to be increased to compensate for a decrease in the other input, capital in this case, while keeping output constant. The MRTS from labor and capital can be derived using the formula

$$MRTS_{L\to K} = \frac{MP_K}{MP_L}$$

where MP_K and MP_L represent respectively the marginal product of capital and the marginal product of Labor. These are the partial derivatives of the production function with respect to capital and labor as seen above. We can then simplify,

$$MRTS_{L\to K} = \frac{\alpha\frac{Y}{K}}{\beta\frac{Y}{L}} = \frac{\alpha}{\beta}\cdot\frac{L}{K}.$$

Let us now take the second derivative of the marginal productivity for each factor. We start with the marginal productivity of capital with respect to capital

$$\frac{\partial MP_K}{\partial K} = \frac{\partial^2 Y}{\partial K^2} = \alpha(\alpha-1)\frac{Y}{K^2}.$$

Similarly, for the marginal productivity of labor with respect to labor:

$$\frac{\partial MP_L}{\partial L} = \frac{\partial^2 Y}{\partial L^2} = \beta(\beta-1)\frac{Y}{L^2}.$$

It is important to note that since both α and β are less than 1, this leads to the fact that $\frac{\partial MP_K}{\partial K}$ and $\frac{\partial MP_L}{\partial L}$ are both negative. In other words, the Cobb-Douglass function satisfies the low of diminishing return, as capital, or labor, increases, the total output increases at a diminishing rate. (Keeping in mind that as one factor increases, we keep the other factor constant).

Example 9.15

The production function is given by $Y = 6K^2 L$.

1. Compute the Marginal Rate of Technical Substitution (MRTS) from K to L.
2. If $K = 4$, how is the production function expressed?
3. What then becomes the value of the MRTS from K to L?

Solution:

1. The Marginal Rate of Technical Substitution (MRTS) of K for L is given by the ratio of the marginal products of L and K. So first we calculate the marginal products

$$\mathrm{MP}_K = \frac{\partial Y}{\partial K} = 12KL,$$

$$\mathrm{MP}_L = \frac{\partial Y}{\partial L} = 6K^2.$$

Thus, the MRTS from L to K is

$$MRTS_{L \to K} = \frac{\mathrm{MP}_L}{\mathrm{MP}_K} = \frac{6K^2}{12KL} = \frac{K}{2L}.$$

2. Substituting $K = 4$ into the production function, we get

$$Y = 6 \times 4^2 \times L = 96L.$$

So the production function becomes $Y = 96L$

3. Since the production function now depends solely on the labor factor L, it is no longer possible to substitute the factor K for the factor L, therefore $MRTS_{K \to L} = 0$.

Example 9.16

A company manufactures a product Y using two production factors: capital (K) and labor (L). The two factors are used interchangeably with a variable rate. Knowing that the production scale or the total factor productivity is equal to one unit, the elasticity of production with respect to labor is $\frac{1}{3}$, and the elasticity of production with respect to capital is also $\frac{1}{3}$.

1. Write the production function.
2. Assume that K is fixed at $K_0 = 6$, interpret this economically. What will then be the amount of labor that the company must employ to achieve a production level of 7 units?
3. Calculate the Marginal Rate of Technical Substitution (MRTS) from L to K and interpret.

Solution:

1. The production function is written as: $Y = AK^\alpha L^\beta$ with $\alpha > 0$, and $\beta > 0$. Since $A = 1$ and $\alpha = \beta = \frac{1}{3}$, therefore $Y(K, L) = K^{1/3}L^{1/3}$.

2. $K = K_0 = 6$ means that capital is the fixed factor. Thus, the produced quantity is independent of the capital factor, and varies only with the labor factor. Consequently, we are in a short-run period. To produce 7 units of output, knowing that capital is fixed at 6 units, the producer must use L units of labor, which is solution to the equation $7 = 6^{1/3}L^{1/3}$. Therefore $L = \frac{343}{6}$.

3. We start with the marginal productivity of K and L

$$MP_K = \frac{\partial Y(K,L)}{\partial K} = \frac{1}{3}K^{-2/3}L^{1/3},$$

$$MP_L = \frac{\partial Y(K,L)}{\partial L} = \frac{1}{3}K^{1/3}L^{-2/3}$$

MRTS from K to L is given by

$$MRTS_{K \to L} = \frac{MP_K}{MP_L} = \left(\frac{\frac{1}{3}K^{-2/3}L^{1/3}}{\frac{1}{3}K^{1/3}L^{-2/3}} \right) = \frac{L}{K}.$$

We notice here that $MRTS$ is a decreasing function in L which means that as more labor is used relative to capital, the rate at which labor can substitute for capital decreases. In simpler terms, adding more labor while keeping capital constant reduces the effectiveness of substituting labor for capital in terms of maintaining the same level of output.

Example 9.17

A firm produces bread using dry yeast and sourdough. Bread production is given by the following function: $Y = F(D, S) = 9D^{1/3}S^{2/3}$, where D and S respectively represent the quantity of dry yeast used and the S the quantity of sourdough used.

1. Find the nature of this firm's returns to scale.
2. Calculate the marginal productivity of the factor D and the sourdough factor S used in the bread production.
3. Calculate and interpret economically the MRTS from D to S.

Solution:

1. To find the nature of the returns to scale for this function F we can find the degree of homogeneity of the production function. Here, as seen above, the production function is of the Cobb-Douglas type. It is therefore homogeneous of degree $\alpha + \beta$. The production function is therefore homogeneous of degree $\frac{1}{3} + \frac{2}{3} = 1$. The returns to scale are therefore constant.

Another method: is to compute the below

$$F(\lambda D, \lambda S) = 9(\lambda D)^{1/3}(\lambda S)^{2/3}$$
$$= 9\lambda^{1/3}\lambda^{2/3}D^{1/3}S^{2/3}$$
$$= \lambda F(D, S).$$

The production function is homogeneous of degree 1, therefore, the returns to scale are constant.

2. The marginal productivity of the factors D and S are given as

$$\text{MP}_D = \frac{\partial F(D, S)}{\partial D} = 9 \times \frac{1}{3}D^{-2/3}S^{2/3},$$
$$\text{MP}_S = \frac{\partial F(D, S)}{\partial S} = 9 \times \frac{2}{3}D^{1/3}S^{-1/3}.$$

3. MRTS from D to S is given by

$$\text{MRTS}_{D \to S} = \frac{\text{MP}_S}{\text{MP}_D} = \left(\frac{9 \times \frac{2}{3}D^{1/3}S^{-1/3}}{9 \times \frac{1}{3}D^{-2/3}S^{2/3}} \right) = \frac{2D}{S}.$$

Therefore the firm must have $\frac{2D}{S}$ additional quantity of factor D to compensate for the loss of one unit of factor S while keeping production at an unchanged level.

9.5 LAGRANGE MULTIPLIER

In many practical applications of maximization or minimization, the problem is to maximize or minimize a given function subject to certain conditions or constraints on the variables involved. The Lagrange method, explained below, applies to any number of variables and constraints and is a powerful tool in economics for finding the extrema of a function subject to constraints. In essence, it allows us to maximize or minimize a function while satisfying a set of conditions on the variables. It is particularly useful in economic analysis when optimizing a function, such as a utility or profit function, subject to certain restrictions, such as a budget constraint or production limitations. A function that is to be either maximized or minimized subject to constraints is often called the **objective function**.

The basic idea of the Lagrange method is to form a new function, called the **Lagrangian**, that includes both the original objective function and the constraints, using a **Lagrange multiplier**. The Lagrange multiplier is a scalar value that is multiplied by the constraints and added to the objective function. The resulting function is then optimized by taking the partial derivatives with respect to all the variables, including the Lagrange multiplier, and setting them equal to zero. We will discuss the procedure, but not the proof, here.

Assume that the objective function, called $f(x, y)$, must be maximized or minimized under the constraint $g(x, y) = 0$. We first form the auxiliary function called a Lagrangian and denoted by $\mathcal{L}$. This Lagrangian function has as variables x, y and the Lagrange multiplier λ;

$$\mathcal{L}(x, y, \lambda) = f(x, y) - \lambda g(x, y).$$

The Lagrange multiplier λ is an unknown scalar.

The solution to the constrained optimization problem can be found by setting the partial derivatives of the Lagrangian with respect to x, y, and λ equal to zero and solving the resulting system of equations:

$$\frac{\partial \mathcal{L}}{\partial x} = \frac{\partial f}{\partial x} - \lambda \frac{\partial g}{\partial x} = 0,$$
$$\frac{\partial \mathcal{L}}{\partial y} = \frac{\partial f}{\partial y} - \lambda \frac{\partial g}{\partial y} = 0,$$
$$\frac{\partial \mathcal{L}}{\partial \lambda} = -g(x, y) = 0.$$

These equations can be rewritten as:

$$\frac{\partial f}{\partial x} = \lambda \frac{\partial g}{\partial x},$$
$$\frac{\partial f}{\partial y} = \lambda \frac{\partial g}{\partial y},$$
$$g(x, y) = 0.$$

These equations are called the **Lagrange conditions** and must be satisfied simultaneously to find the maximum or minimum value of $f(x, y)$ subject to the constraint $g(x, y)$. Solving this system of equations will give us the critical points of $\mathcal{L}(x, y, \lambda)$ subject to the constraint $g(x, y) = 0$. Note that the third equation is simply the constraint equation $g(x, y) = 0$.

The significance of the Lagrange multiplier, λ, extends beyond its mathematical role. Notice that

$$\lambda = \frac{\partial f / \partial x}{\partial g / \partial x} = \frac{\partial f / \partial y}{\partial g / \partial y} = \frac{\partial f}{\partial g}.$$

Thus, λ can be interpreted as the rate of change of the objective function $f(x, y)$ with respect to a small change in the constraint. In other words, it quantifies the marginal utility or benefit of changing the constraint by one unit. This interpretation is particularly useful in economic analysis, where understanding the trade-offs between different objectives is crucial.

The solution of the system of equations with the unknowns x, y, and λ above provides the critical points of the Lagrangian function subject to the constraint. These critical points satisfy the constraint, but to determine whether they represent an extremum of the original objective function f, we employ a modified second derivative test tailored for the context of Lagrange multipliers. Specifically, we examine the second derivatives of the Lagrangian $\mathcal{L}$, not just f. To do so, we follow the same process followed in the previous section as part of the second derivative test. Let

$$D = \frac{\partial^2 \mathcal{L}}{\partial x^2} \times \frac{\partial^2 \mathcal{L}}{\partial y^2} - \left(\frac{\partial^2 \mathcal{L}}{\partial x \partial y} \right)^2.$$

- We have a maximum at (x_0, y_0) if $D > 0$, $\frac{\partial^2 \mathcal{L}}{\partial x^2} < 0$ and $\frac{\partial^2 \mathcal{L}}{\partial y^2} < 0$.
- We have a minimum at (x_0, y_0) if $D > 0$, $\frac{\partial^2 \mathcal{L}}{\partial x^2} > 0$ and $\frac{\partial^2 \mathcal{L}}{\partial y^2} > 0$.
- If $D < 0$, the critical point is a saddle point, and thus neither a local maximum nor a local minimum.

In order to find global extrema on the domain of x and y it would also be necessary to examine the function on the boundaries of the domain, in a manner reminiscent of the Extreme Value Theorem.

Example 9.18

Maximize the function $f(x, y) = xy$ subject to the constraint $x^2 + y^2 = 1$.

Solution: Our constraint condition gives us $g(x, y) = x^2 + y^2 - 1 = 0$. We can use the Lagrange method to solve this problem by considering the Lagrange function

$$\mathcal{L}(x, y, \lambda) = xy - \lambda(x^2 + y^2 - 1).$$

The system of equations we need to solve is:

$$\frac{\partial \mathcal{L}}{\partial x} = y - 2\lambda x = 0,$$

$$\frac{\partial \mathcal{L}}{\partial y} = x - 2\lambda y = 0,$$

$$\frac{\partial \mathcal{L}}{\partial \lambda} = x^2 + y^2 - 1 = 0.$$

From the first equation, we get $y = 2\lambda x$, and from the second equation, we get $x = 2\lambda y$. Substituting these into the third equation, we get $\lambda^2 = \frac{1}{4}$. This gives us the four critical points

$$\left(\frac{1}{\sqrt{2}}, \frac{1}{\sqrt{2}} \right), \left(-\frac{1}{\sqrt{2}}, -\frac{1}{\sqrt{2}} \right), \left(-\frac{1}{\sqrt{2}}, \frac{1}{\sqrt{2}} \right), \left(\frac{1}{\sqrt{2}}, -\frac{1}{\sqrt{2}} \right).$$

To determine whether these critical points correspond to a maximum, minimum, or saddle point, we need to examine the second partial derivatives of $\mathcal{L}(x, y, \lambda)$.

$$D = \frac{\partial^2 \mathcal{L}}{\partial x^2} \times \frac{\partial^2 \mathcal{L}}{\partial y^2} - \left(\frac{\partial^2 \mathcal{L}}{\partial x \partial y}\right)^2 = (-2\lambda)(-2\lambda) - (1)(1) = 4\lambda^2 - 1$$

Since we found $\lambda = \pm\frac{1}{2}$, substituting these values into the determinant gives us $D = 4\left(\frac{1}{2}\right)^2 - 1 = 0$. This indicates that our test is inconclusive at this step, suggesting that the critical points could be saddle points, but further investigation is required to conclusively determine their nature.

Given the constraint $x^2 + y^2 = 1$ and the function $f(x, y) = xy$, we can reason about the nature of these points. Since the constraint describes a circle and we are maximizing the product xy, the points $\left(\frac{1}{\sqrt{2}}, \frac{1}{\sqrt{2}}\right)$ and its variants across quadrants represent the maximum and minimum values of $f(x, y)$ on the unit circle. These points are thus the maxima and minima of f subject to the given constraint.

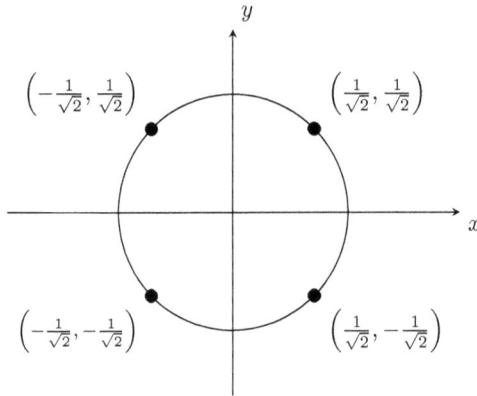

We evaluate the function $f(x, y) = xy$ at each of the critical points to confirm the maximum and minimum values.

$$f\left(\frac{1}{\sqrt{2}}, \frac{1}{\sqrt{2}}\right) = \frac{1}{\sqrt{2}} \cdot \frac{1}{\sqrt{2}} = \frac{1}{2}$$

and

$$f\left(-\frac{1}{\sqrt{2}}, -\frac{1}{\sqrt{2}}\right) = \left(-\frac{1}{\sqrt{2}}\right)\left(-\frac{1}{\sqrt{2}}\right) = \frac{1}{2}$$

Thus, the maximum value of $f(x, y)$ subject to the constraint $g(x, y)$ is $\frac{1}{2}$, which is achieved at the points $\left(\frac{1}{\sqrt{2}}, \frac{1}{\sqrt{2}}\right)$ and $\left(-\frac{1}{\sqrt{2}}, -\frac{1}{\sqrt{2}}\right)$.

On the other hand,

$$f\left(\frac{1}{\sqrt{2}}, -\frac{1}{\sqrt{2}}\right) = \frac{1}{\sqrt{2}} \cdot \left(-\frac{1}{\sqrt{2}}\right) = -\frac{1}{2}$$

and

$$f\left(-\frac{1}{\sqrt{2}}, \frac{1}{\sqrt{2}}\right) = \left(-\frac{1}{\sqrt{2}}\right)\frac{1}{\sqrt{2}} = -\frac{1}{2}.$$

Thus, the minimum value of $f(x,y)$ subject to the constraint $g(x,y)$ is $-\frac{1}{2}$, which is achieved at the points $\left(\frac{1}{\sqrt{2}}, -\frac{1}{\sqrt{2}}\right)$ and $\left(-\frac{1}{\sqrt{2}}, \frac{1}{\sqrt{2}}\right)$.

Example 9.19

Determine the minimum and maximum values of the objective function $f(x, y) = 5x^2 + 6y^2 - xy$ subject to the constraint that $x + 2y = 24$.

Solution: We can use the Lagrange method to solve this problem by considering the Lagrange function

$$\mathcal{L}(x, y, \lambda) = 5x^2 + 6y^2 - xy - \lambda(x + 2y - 24).$$

The system of equations we need to solve is:

$$\frac{\partial \mathcal{L}}{\partial x} = 10x - y - \lambda = 0,$$

$$\frac{\partial \mathcal{L}}{\partial y} = 12y - x - 2\lambda = 0,$$

$$\frac{\partial \mathcal{L}}{\partial \lambda} = x + 2y - 24 = 0.$$

From the first two equations, we get $y = \frac{3}{2}x$, and from the third equation we get $x = 6$, and $y = 9$, giving us the critical point $(6, 9)$.

To find the minimum value, we can use the second derivative test to confirm that the critical point we found is a local minimum. Taking the second partial derivatives of $\mathcal{L}(x, y)$ we get:

$$\frac{\partial^2 \mathcal{L}}{\partial x^2} = 10,$$

$$\frac{\partial^2 \mathcal{L}}{\partial y^2} = 12,$$

$$\frac{\partial^2 \mathcal{L}}{\partial x \partial y} = -1.$$

We then compute

$$D = \frac{\partial^2 \mathcal{L}}{\partial x^2} \times \frac{\partial^2 \mathcal{L}}{\partial y^2} - \left(\frac{\partial^2 \mathcal{L}}{\partial x \partial y}\right)^2 = (10)(12) - (-1)^2 = 119.$$

Since this is positive and both $\frac{\partial^2 \mathcal{L}}{\partial x^2} > 0$ and $\frac{\partial^2 \mathcal{L}}{\partial y^2} > 0$, the critical point is a local minimum. Therefore, the minimum value of the objective function is:

$$f(6,9) = 5(6)^2 + 6(9)^2 - 6(9) = 612.$$

Example 9.20

Consider the function $f(x,y) = 3x^2 - 4xy + 2y^2$. We want to find the maximum and minimum values of f subject to the constraint $x + y = 10$.

Solution: We can use the Lagrange method to solve this problem by considering the Lagrange function

$$\mathcal{L}(x,y,\lambda) = 3x^2 - 4xy + 2y^2 - \lambda(x + y - 10).$$

The system of equations we need to solve is:

$$\frac{\partial \mathcal{L}}{\partial x} = 6x - 4y - \lambda = 0,$$

$$\frac{\partial \mathcal{L}}{\partial y} = 4y - 4x - \lambda = 0,$$

$$\frac{\partial \mathcal{L}}{\partial \lambda} = x + y - 10 = 0.$$

From the first two equations, we get $x = \frac{4}{5}y$, and from the third equation, we get $y = \frac{50}{9}$. We get $(\frac{40}{9}, \frac{50}{9})$, which is the only critical point.

To determine whether this critical point is a maximum or a minimum, we need to use the second partial derivative test. We calculate the second partial derivatives of $\mathcal{L}$:

$$\frac{\partial^2 \mathcal{L}}{\partial x^2} = 6,$$

$$\frac{\partial^2 \mathcal{L}}{\partial y^2} = 4,$$

$$\frac{\partial^x \mathcal{L}}{\partial x \partial y} = -4.$$

We then compute

$$D = \frac{\partial^2 \mathcal{L}}{\partial x^2} \times \frac{\partial^2 \mathcal{L}}{\partial y^2} - \left(\frac{\partial^2 \mathcal{L}}{\partial x \partial y} \right)^2 = (6)(4) - (-4)^2 = 8.$$

Since this is positive and both $\frac{\partial^2 \mathcal{L}}{\partial x^2} > 0$ and $\frac{\partial^2 \mathcal{L}}{\partial y^2} > 0$, the critical point is a local minimum. Therefore, the minimum value of the objective function is

$$f \left(\frac{40}{9}, \frac{50}{9} \right) = \frac{200}{9} \approx 22.22.$$

9.6 ECONOMIC APPLICATIONS

There are many economic applications of constrained minima and maxima. For example, if a producer produces two goods, they may want to minimize the total cost while still having to produce a specified minimum total quantity. A company may want to maximize sales resulting from two advertisements while adhering to the advertising budget constraint. A consumer may want to maximize their utility function by consuming certain goods while being restricted by their budget.

Example 9.21

A company's daily budget is 48 thousand dollars to buying two goods x and y needed for its operation. The prices of x and y are \$2000 and \$3000 respectively. The company's utility function for these two goods is given by the formula $U = -x^2 - 2y^2 + 2xy$. How many units of good x and y must he consume to maximize his utility?

Solution: The objective is to maximize the utility function $U = -x^2 - 2y^2 + 2xy$ subject to the constraint $2000x + 3000y = 48000$. Dividing the equation by 1000 and rearranging gives $2x + 3y - 48 = 0$. We use these to find the Lagrange function

$$\mathcal{L}(x, y, \lambda) = -x^2 - 2y^2 + 2xy - \lambda(2x + 3y - 48),$$

which in turn gives us the system of equations we need to solve:

$$\frac{\partial \mathcal{L}}{\partial x} = -2x + 2y - 2\lambda = 0,$$

$$\frac{\partial \mathcal{L}}{\partial y} = -4y + 2x - 3\lambda = 0,$$

$$\frac{\partial \mathcal{L}}{\partial \lambda} = 2x + 3y - 48 = 0.$$

Solving for x and y, we get $x = \frac{336}{29}$ and $y = \frac{240}{29}$, which is the only critical point.

To determine whether this critical point is a maximum or a minimum, we need to use the second partial derivative test. We calculate the second partial derivatives of $\mathcal{L}$:

$$\frac{\partial^2 \mathcal{L}}{\partial x^2} = -2,$$

$$\frac{\partial^2 \mathcal{L}}{\partial y^2} = -4,$$

$$\frac{\partial^2 \mathcal{L}}{\partial x \partial y} = 2.$$

We then compute

$$D = \frac{\partial^2 \mathcal{L}}{\partial x^2} \times \frac{\partial^2 \mathcal{L}}{\partial y^2} - \left(\frac{\partial^2 \mathcal{L}}{\partial x \partial y} \right)^2 = (-2)(-4) - (2)^2 = 4$$

Since this is positive and both $\frac{\partial^2 \mathcal{L}}{\partial x^2} < 0$ and $\frac{\partial^2 \mathcal{L}}{\partial y^2} < 0$, the critical point is a local maximum.

To maximize the consumer's utility given the utility function $U = -x^2 - 2y^2 + 2xy$ and the budget constraint $48 = 2x + 3y$, the consumer should consume $\frac{336}{29}$ units of good x and $\frac{240}{29}$ units of good y.

Example 9.22

A firm produces devices in two different factories. The total production costs for the two factories are, respectively,

$$CT_1 = 200 + 6q_1 + 0.03q_1^2,$$
$$CT_2 = 150 + 10q_2 + 0.02q_2^2,$$

where q_1 and q_2 represent the number of devices produced in each factory. The firm has committed to delivering 100 devices to a company. Transportation costs per device are \$4 for deliveries from the first factory and \$2 for deliveries from the second factory. Transport costs are borne by the productive firm. Calculate the number of devices that the firm must produce in each factory to minimize the total cost of production including the cost of transportation.

Solution: Let us find the number of devices that the firm must produce in each factory in order to minimize the total cost of production including the cost of transport. The total cost is equal to:

$$TC = (CT_1 + 4q_1) + (CT_2 + 2q_2)$$
$$= 200 + 6q_1 + 0.03q_1^2 + 4q_1 + 150 + 10q_2 + 0.02q_2^2 + 2q_2$$
$$= 0.03q_1^2 + 0.02q_2^2 + 10q_1 + 12q_2 + 350.$$

Therefore, it is a matter of minimizing this function, and this is under the constraint of delivering 100 devices in total, $q_1 + q_2 = 100$. The Lagrange function in this case is given as

$$\mathcal{L}(q_1, q_2, \lambda) = 0.03q_1^2 + 0.02q_2^2 + 10q_1 + 12q_2 + 350 - \lambda(q_1 + q_2 - 100).$$

As we have done in the previous example, the system of equations we need to solve is:

$$\frac{\partial \mathcal{L}}{\partial q_1} = 0.06q_1 + 10 - \lambda = 0,$$

$$\frac{\partial \mathcal{L}}{\partial q_2} = 0.04q_2 + 12 - \lambda = 0,$$

$$\frac{\partial \mathcal{L}}{\partial \lambda} = q_1 + q_2 - 100 = 0.$$

Solving for q_1 and q_2, we find $q_1 = 60$ and $q_2 = 40$. It is still necessary to verify that, for these values, it is indeed a minimum. To do so we need to use the second partial derivative test. We calculate the second partial derivatives of $\mathcal{L}$:

$$\frac{\partial^2 \mathcal{L}}{\partial q_1^2} = 0.06,$$

$$\frac{\partial^2 \mathcal{L}}{\partial q_2^2} = 0.04,$$

$$\frac{\partial^2 \mathcal{L}}{\partial x \partial y} = 0.$$

We then compute

$$D = \frac{\partial^2 \mathcal{L}}{\partial q_1^2} \times \frac{\partial^2 \mathcal{L}}{\partial q_2^2} - \left(\frac{\partial^2 \mathcal{L}}{\partial q_1 \partial q_2}\right)^2 = (0.06)(0.04) - 0 = 0.0024.$$

Since D, $\frac{\partial^2 \mathcal{L}}{\partial q_1^2}$, and $\frac{\partial^2 \mathcal{L}}{\partial q_2^2}$ are all positive, the critical point is a local minimum. Therefore, when the firm delivers 60 devices from its first factory and 40 from its second factory, the total cost is minimal under the constraint of a delivery of 100 devices.

Example 9.23

Consider a company that has a budget of $100 (thousand dollars) for investing in two types of advertising: online (x) and television (y). The cost per unit of online and television advertising is $5 and $20, respectively. The company's revenue function, which depends on the amount spent on these two types of advertising, is given by: $R(x, y) = 4x^{0.5}y^{0.5}$. How much should the company invest in each type of advertising to maximize its revenue?

Solution: The objective is to maximize the revenue function $R = 4x^{0.5}y^{0.5}$ subject to the constraint $5x + 20y = 100$, or $5x + 20y - 100 = 0$. We formulate the Lagrange function:

$$\mathcal{L}(x, y, \lambda) = 4x^{0.5}y^{0.5} - \lambda(5x + 20y - 100)$$

The system of equations we need to solve is derived from the partial derivatives of $\mathcal{L}$ with respect to x, y, and λ:

$$\frac{\partial \mathcal{L}}{\partial x} = 2y^{0.5}x^{-0.5} - 5\lambda = 0,$$

$$\frac{\partial \mathcal{L}}{\partial y} = 2x^{0.5}y^{-0.5} - 20\lambda = 0,$$

$$\frac{\partial \mathcal{L}}{\partial \lambda} = 5x + 20y - 100 = 0.$$

Solving this system of equations allows us to find the values of x, y, and λ that maximize the company's revenue under the given budget constraint.

From the first two equations, we can express λ in terms of x and y:

$$\lambda = \frac{2y^{0.5}}{5x^{0.5}},$$

$$\lambda = \frac{2x^{0.5}}{20y^{0.5}}.$$

Equating the two expressions for λ gives us a $4y = x$. Substituting this relationship and the budget constraint into the third equation allows us to solve to find $x = 10$ and $y = \frac{10}{4}$. Therefore, the optimal investment strategy involves allocating 10 (thousand dollars) of the budget to online advertising and the remainder to television advertising, ensuring the maximum possible revenue within the budget constraints.

9.7 PROBLEMS

Question 9.1 *Use Implicit differentiation to find y':*

(a) $y^3 - 3x^2 - 6x - 1 = 0$

(b) $\frac{2}{3}y^3 - 6y - 3x^2 = 4$

(c) $y^2 + x^2 = 25$

(d) $6xy - 2x^3 - 5 = 0$

(e) $3x^3y - 6y^2 - 5x = 2$

(f) $x - \frac{1}{2}y^2x - 5y = 6x^2$

(g) $1 - x^3 - 3y^2 = 0$

(h) $\frac{x^2-y}{6x} = 1$

(i) $4\sqrt{x+y} - 5x^2 - 1 = 0$

Question 9.2 *Find the following limits:*

(a) $\lim_{x \to -4} \frac{4-x^2}{x+4}$

(b) $\lim_{x \to 0} \frac{x^3-3x}{2x}$

(c) $\lim_{x \to 2} \frac{x^2-5x+6}{x-2}$

(d) $\lim_{x \to -1} \frac{x+1}{x^3+1}$

(e) $\lim_{x \to -2} \frac{(x+2)^4}{x+2}$

(f) $\lim_{x \to \infty} \frac{(2x^3-4x+2)}{x^4+2x^2-5x}$

Question 9.3 *Determine whether demand is inelastic, elastic, or unit elastic at $p = 10$ for each of the following price-demand functions. (Find $\epsilon(10)$.)*

(a) $120 - 2x - p = 0$

(b) $x = 40 - 2.5p$

(c) $3p + x = 50$

(d) $3x + 5p = 120$

Question 9.4 *Given the price-demand function: $50p + 10x - 600 = 0$.*

(a) *Find the demand x as a function of the price p.*

(b) *Find the elasticity of demand when $p = 5\$$. If this price increases by 10%, what is the new elasticity of demand?*

(c) *Find the elasticity of demand when $p = 8\$$.*

(d) *If this price increases by 10%, what is the new elasticity of demand?*

Question 9.5 *Given the price-demand function: $0.4x + p = 20$.*

(a) *Write the demand x as a function of the price p.*

(b) *Write the revenue as a function of the price p.*

(c) *Find the elasticity of demand $\epsilon(p)$.*

(d) *For which price p is the demand elastic or inelastic?*

(e) *On which intervals is revenue increasing and/or decreasing?*

Question 9.6 *Consider a cheese factory that uses milk and rennet as primary inputs. The production function for cheese is given by: $Y = F(M, R) = 12M^{1/2}R^{1/2}$, where M represents the quantity of milk used, and R represents the quantity of rennet used.*

(a) *Determine the nature of the factory's returns to scale.*

(b) *Calculate the marginal productivity of the factors M (milk) and R (rennet).*

(c) *Calculate and interpret economically the Marginal Rate of Technical Substitution (MRTS) from M to R.*

Question 9.7 *Consider a manufacturing firm that uses capital and labor to produce gadgets. The production function for the firm is represented by the Cobb-Douglas function: $Y = AK^{0.4}L^{0.6}$ where Y is the total output, K is the amount of capital used, L is the amount of labor used, and A is a constant factor of total factor productivity.*

   (a) Calculate the marginal productivity of labor MP_L.

   (b) Calculate the marginal productivity of capital MP_K.

   (c) Discuss how an increase in labor affects the marginal productivity of capital and vice versa.

Question 9.8 *Find the maximum and minimum values of $f(x,y) = x^2 + y^2$ subject to the constraint $x + y = 1$.*

Question 9.9 *Find the maximum and minimum values of $f(x,y) = x^2y$ subject to the constraint $x^2 + y^2 = 1$.*

Question 9.10 *A firm faces a production constraint given by $F(x,y) = x^2 + y^2 \le 25$, and wants to minimize its cost of production, $C(x,y) = ax + by$, where a and b are positive constants representing the costs of producing x and y, respectively. What is the firm's cost-minimizing combination of x and y that satisfies the production constraint?*

Question 9.11 *A firm produces two types of goods, x and y, and has a production function given by $F(x,y) = 5x + 3y$. The production process, however, is constrained by the technology available, which dictates that the combined production of both products cannot exceed a maximum capacity, represented by $x + y \le 100$. Find the profit-maximizing levels of x and y assuming there are no other costs involved. If the business decides to produce only one type of product due to market demand, determine which product should be produced to maximize profit, and calculate the maximum possible profit.*

Question 9.12 *A company manufactures two types of devices x and y. The function of the joint cost is given as: $f(x,y) = x^2 + 2y^2 - xy$. How many devices should each type of company manufacture to minimize its cost if it needs a total of 8 devices?*

Question 9.13 *A consumer spends his daily income of $90 on the purchase of two goods x and y, priced at $5 and $10 respectively. The consumer's utility function is given by: $U = -x^2 - 2y^2 + xy$. How many units of goods x and y are needed to maximize his utility?*

Question 9.14 *A consumer has a utility function $U(x,y) = x^\alpha y^\beta$, where α and β are positive constants, and faces a budget constraint given by $p_1x + p_2y = M$, where p_1 and p_2 are the prices of goods x and y, respectively, and M is the consumer's income. What are the utility-maximizing values of x and y?*

Answers to Questions

A.1   ANSWERS TO CHAPTER 1

Question 1.1:
(a) True   (b) False   (c) True

Question 1.2: $\frac{1}{2}$

Question 1.3:
(a) $-3x^2 + 14x - 8$ (d) $79p^2 + 108p - 20q - 20$

(b) $5x^2 - 25x + 35$ (e) $27x^3 + 135x^2 + 225x + 125$

(c) $6 - 2x^4 + 4x^2$ (f) $4x^4 - 4x^3 + 13x^2 - 6x + 9$

Question 1.4: $f(x) = x^2 + 3x - 1, f(h+2) = h^2 + 7h + 9$

Question 1.5: $f(p) = p(p - 10), f(10) = 0$

Question 1.6: $f(x) = \frac{1}{3}x + 2$ and $g(x) = x^2 - 3$. $f \circ g(x) = \frac{1}{3}x + 1$.

Question 1.7: $f(x) = \frac{x+2}{3x-1}$, and $g(x) = x^3 + 2x - 1$. $f \circ g(1) = \frac{4}{5}$

Question 1.8:
(a) $x(x - 10)$ (e) $x(x - 3\sqrt{3})(x + $ (h) $(3x - 2)(7x - 3)$
(b) $x(x^2 - 3x + 6)$ $3\sqrt{3})$ (i) $-(3x - 2)(x - 4)$
(c) $(x - 2)(x + 3)$ (f) $(x + 1)(x + 4)$
(d) $x(x - 2)(x - 3)$ (g) $(2x - 1)(x - 1)$

Question 1.9:
(a) $x = \frac{2}{3}$ or $x = 4$ (d) $x = 3$
(b) $x = \frac{5}{3}$ or $x = 2$ or $x = 3$ (e) $x = -4$ or $x = -1$
(c) $x = 0$ or $x = 2$ or $x = 3$ (f) $x = \frac{3}{7}$ or $x = \frac{2}{3}$

Question 1.10:

(a) $\mathbb{R}$ (all real numbers)   (d) $x \neq \pm 3$
(b) $x \neq \pm 4$ (e) $\{x \in R : x \geq 4\}$
(c) $\mathbb{R}$ (all real numbers)   (f) $\mathbb{R}$ (all real numbers)

DOI: 10.1201/9781003480235-A

Question 1.11:

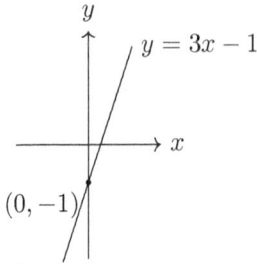

$y = 3x - 1$

$(0, -1)$

Question 1.12:

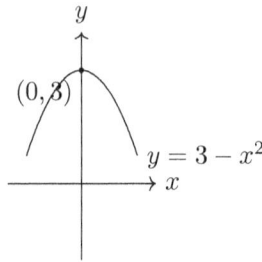

$(0, 3)$

$y = 3 - x^2$

Question 1.13:

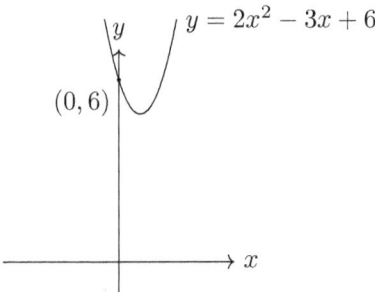

$y = 2x^2 - 3x + 6$

$(0, 6)$

Question 1.14: $D : p = -\frac{15}{8}x + 150$

Question 1.15: $D : p = -\frac{1}{4}x + \frac{54}{4}$

Question 1.16: $D : p = \frac{1}{2}x + 3$

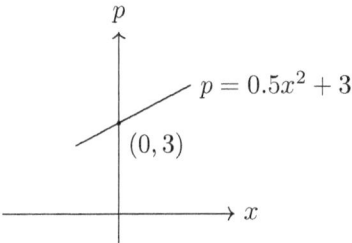

$p = 0.5x^2 + 3$

$(0, 3)$

Question 1.17: Equilibrium point= $(2, 4)$, Supply= 2 and Price= \$4

Question 1.18: Supply= 1 and Price=\$6

Question 1.19: Supply= 20 and Price=\$9

Question 1.20: Demand: $p = -2x + 9$. To find the exact supply function, we need another point or additional information about its slope or intercept. Without this, the supply function remains undefined beyond the fact that it passes through the point $(2, 5)$.

Question 1.21: Demand: $p = -3x + 21$. To find the exact supply function, we need another point or additional information about its slope or intercept. Without this, the supply function remains undefined beyond the fact that it passes through the point $(4, 9)$.

Question 1.22: $a = \frac{-1}{9}$ and $y_0 = 14$

Question 1.23:

(a) The slope $\frac{-2}{3}$ means for a three unit increase in quantity, price falls two dollars.

(b) The p-intercept 120 means when the price is $120, the quantity demanded is zero.

(c) The demand is 165 when the price is 10.

(d) Graph the demand function for $0 < x < 150$.

(e) $x = D(p) = 180 - \frac{3}{2}p$.

Question 1.24

(a) The slope $\frac{1}{2}$ means for a one dollar increase in price, supply increases two units.

(b) The p-intercept 120 means for a price of $120, no units are produced.

(c) The supply is 160 when the price is $200.

(d) Graph the supply function for $0 < x < 150$.

(e) $x = D(p) = 2p - 240$.

Question 1.25: $TC = 2500 + 20x$

Question 1.26:

(a) $TC(x) = 1500 + 9x$

(b) 540

(c) 108 meals

(d) $TR(x) = 12x$

(e) 10 meals

(f) Profit$= 12x - (1500 + 9x) = 3x - 1500$

(g) $900

Question 1.27:

(a) $TR(x) = 85x$

(b) 25 suitcases

(c) $px = 2000 + 5x$

(d) $p = \frac{130}{6}$

A.2   ANSWERS TO CHAPTER 2

Question 2.1:

(a) 40   (b) 130   (c) 100

Question 2.2:

(a) 60   (b) 0   (c) −60

Question 2.3: 3

Question 2.4: 59

Question 2.5: $4x$

Question 2.6: $-100x$

Question 2.7:

(a) 27   (b) 75   (c) 147

Question 2.8:

(a) 7   (b) 448   (c) 5103

Question 2.9:

(a) 1   (b) 1   (c) 1

Question 2.10:

(a) 6   (b) 30   (c) 36

Question 2.11:

(a) 0   (b) 160   (c) 540

Question 2.12:

(a) −6   (b) −54   (c) −294

Question 2.13:

(a) −2   (b) $\frac{-1}{4}$ (c) $\frac{-2}{125}$

Question 2.14:

(a) −5   (b) $\frac{-5}{729}$ (c) $\frac{-1}{3125}$

Question 2.15:

(a) $\frac{-15}{16}$ (b) $\frac{-15}{256}$ (c) $\frac{-15}{625}$

Question 2.16: Does not exist

Question 2.17: $\frac{-3}{2}$

Question 2.18: Does not exist

Question 2.19: $F'(x) = 3x^2$, $F'(6) = 108$

Question 2.20: $\frac{1}{10}$

Question 2.21: $70x^4 + 21x^2 - 10$

Question 2.22: $\frac{-27}{x^4} - \frac{24}{x^3} + \frac{10}{x^2}$

Question 2.23: 2135

Question 2.24: −11970

Question 2.25: 38

Question 2.26: 1120

Question 2.27: $8x^7$

Question 2.28: $\frac{1}{4}$

Question 2.29: $\frac{-1}{2\sqrt{x^3}}$

Question 2.30: $\frac{1}{4\sqrt{x}}$

Question 2.31: $3q^2$

Question 2.32: $\frac{3x^2}{4} - 10x + 2$

Question 2.33: $340 - 40x$

Question 2.34: $N(22) = 66 + \frac{1}{\sqrt{22^3}}$ and $N'(22) = 3 - \frac{3}{2\sqrt{22^5}}$

Question 2.35: $\frac{-23}{2\sqrt{10}}$

Question 2.36:

1. (a) $TR = 90q - \sqrt{q^7}$

   (b) $MR = 90 - \frac{7\sqrt{q^5}}{2}$

   (c) $AR = 90 - \sqrt{q^5}$

2. Slope of MR is $\frac{7}{2}$ times that of AR.

3. $90^{\frac{2}{5}}$

4. $\left(\frac{180}{7}\right)^{\frac{2}{5}}$

Question 2.37:

1. $q^3 - 7q^2 + 120 + 60q$

2. 2820

3. 120

4. $q^3 - 7q^2 + 60q$

5. $3q^2 - 14q + 60$

A.3   ANSWERS TO CHAPTER 3

Question 3.1:

(a) $f'(x) = \frac{1}{15}$

(b) $f'(x) = \frac{15}{x}$

(c) $f'(x) = \frac{8}{\ln(2)x}$

(d) $f'(x) = \frac{10}{\ln(7)x} + 2x$

(e) $f'(x) = \ln(30)30^x$

(f) $f'(x) = e^x + 10$

(g) $f'(x) = 3\ln(2)2^x - 4e^x$

(h) $f'(x) = \frac{1}{\ln(1/7)x}$

(i) $f'(x) = \ln\left(\frac{3}{5}\right)\left(\frac{3}{5}\right)^x$

Question 3.2: 23.45 years (Rounded to the nearest hundredth).

Question 3.3: 3% interest rate compound every 6 months.

Question 3.4: 7112.83 (Rounded to the nearest hundredth).

Question 3.5:

(a) 0.51 $mmHg$ per kg (Rounded to the nearest hundredth).

(b) 64.71kg (Rounded to the nearest hundredth).

Question 3.6:

1. Exponential growth.
2. Exponential decay.
3. Exponential growth.

Question 3.7:$P(t) = 60(2)^{\frac{3t}{10}}$

Question 3.8: -32.30 and -2.93 micrograms per milliliter (Rounded to the nearest hundredth).

Question 3.9: 1.4 million, 10,467 people per year.

Question 3.10: 10 cases (rounded from 9.99).

Question 3.11: The bacteria population at the beginning was 1 and the rate of change of the bacteria population after 4 years is 1.675 (To the nearest thousand).

Question 3.12: $M = 243.38$

Question 3.13: 4.88%

Question 3.14: 1.48

Question 3.15: 754.48

Question 3.16: Approximately 2.34 thousand units

Question 3.17: -0.06. The rate of change of price with respect to demand at 100 cups is approximately -0.06 dollars per cup, which means that when 100 cups of coffee have been sold, a decrease in price per cup of about 6 cents will result in an additional cup being sold.

Question 3.18:

(a) 261.41

(b) An annual increase rate of 30 books.

Question 3.19:

(a) $\dfrac{dD}{dp} = -8e^{-0.1p}$.

(b) As the price increases, the demand decreases.

A.4   ANSWERS TO CHAPTER 4

Question 4.1

(a) $f'(x) = \frac{(x^2+6)(3x^2)-(x^3+5)(2x+6)}{(x^2+6x)^2}$

(b) $f'(x) = \frac{(x^3-3)(-2x-1)-(-x^2-x)(3x^2)}{(x^3-3)^2}$

(c) $f'(x) = \frac{(x^4-x^3)e^x-e^x(4x^3-3x^2)}{(x^4-x^3)^2}$

(d) $f'(x) = \frac{(x-1)5x^4-x^5}{(x-1)^2}$

(e) $f'(x) = \frac{(x^2-2x)(1/x)-\ln(x)(2x-2)}{(x^2-2x)^2}$

(f) $f'(x) = \frac{\ln(x)\ln(2)2^x-2^x(1/x)}{\left(\ln(x)\right)^2}$

(g) $f'(x) =$

$\quad \frac{(x^2-e^x)(\ln(5)5^x+1)-(5^x+x)(2x-e^x)}{(x^2-e^x)^2}$

(h) $f'(x) =$

$\quad \frac{(-2x^3+5x)(15x^4-6x^2)-(3x^5-2x^3)(-6x^2+5)}{(-2x^3+5x)^2}$

(i) $f'(x) = \frac{2^x\left(\frac{1}{\ln(5)x}\right)-\log_5(x)\ln(2)2^x}{(2^x)^2}$

Question 4.2

(a) $f'(x) = \ln(x) + \frac{x+7}{x}$

(b) $f'(x) = \frac{3^x}{x} + \ln(3)\ln(x)3^x$

(c) $f'(x) = \ln(4)4^x(3x-5) + 3(4^x)$

(d) $f'(x) = 6x^2e^x + 2x^3e^x$

(e) $f'(x) = 30x^5\ln(x) + 5x^6\frac{1}{x}$

(f) $f'(x) = (10x^4 + 12x^3)\ln(x) + (2x^5 + 3x^4)\frac{1}{x}$

(g) $f'(x) = 10(5x^3 - 4x) + 10x(15x^2 - 4)$

(h) $f'(x) = (10x-3)e^x+(5x^2-3x+7)e^x$

(i) $f'(x) = \ln(5)5^x\ln(x) + 5^x\frac{1}{x}$

(j) $f'(x) = e^x7^x + e^x\ln(7)7^x$

(k) $f'(x) = (3x^2 - 4x)2^x + (x^3 - 2x^2)\ln(2)2^x$

(l) $f'(x) = \ln(5)5^x7^x + 5^x\ln(7)7^x$

Question 4.3

(a) $f'(x) = \ln(5)5^{7x}(7)$

(b) $f'(x) = -15e^{3x}$

(c) $f'(x) = e^{4x}$

(d) $f'(x) = \frac{1}{\ln(7)x}$

(e) $f'(x) = \ln(7)7^{x^4}(4x^3)$

(f) $f'(x) = \frac{2}{2x-7}$

(g) $f'(x) = e^{-3x+5}(-3)$

(h) $f'(x) = \frac{3x^2-12x}{x^3-6x^2}$

(i) $f'(x) = 5(3x-5)^4(3)$

(j) $f'(x) = 3(2x^2 + 6x)^2(4x + 6)$

(k) $f'(x) = 2(5x^3-7x^2)(15x^2-14x)$

(l) $f'(x) = 3(-4x^4 + 7x^3 - 6x^2)^2(-16x^3 + 21x^2 - 12x)$

Question 4.4

(a) $f'(x) = \frac{6x-5}{\ln(4)(3x^2-5x+2)}$

(b) $g'(x) = \frac{3x^3-5xe^x+2}{x\ln(10)(x^3-5e^x+2\ln(x))}$

(c) $g'(x) = 3x^2 - (2x+3)\frac{1}{x^2+3x-1}$

(d) $f'(x) = -\frac{3}{x}$

(e) $f'(x) = (6\ln(x^2)-2x+4)(\frac{6}{x}-1)$

(f) $f'(x) = -\frac{3(xe^x-e^x+1)}{(e^x-1)^2}$

Question 4.5

(a) $f'(x) = \frac{4\log(x^2)}{\ln(10)x}$

$\quad f''(x) = \frac{8-4\ln(10)\log(x^2)}{(\ln(10)x)^2}$

(b) $g'(x) = \frac{-e^x}{(e^x-1)^2}$

$\quad g''(x) = \frac{(e^x-1)^2(-e^x)-(-e^x)2(e^x-1)e^x}{(ex-1)^4}$

(c) $g'(x) = \frac{2}{\ln(3)x}$

$\quad g''(x) = \frac{-2}{\ln(3)x^2}$

Question 4.6: 0.13 (Rounded to the nearest hundredth.)

Question 4.7: 0.007 (Rounded to the nearest thousandth.)

Question 4.8: 0.17 (Rounded to the nearest hundredth.)

Question 4.9: 1.5625 .

Question 4.10: 1.543 (Rounded to the nearest thousandth.)

Question 4.11: The positive critical point, which represents the quantity that minimizes the total cost, is approximately 273.86

Question 4.12: The positive critical point, which represents the quantity that minimizes the total cost, is approximately 46.64.

Question 4.13: $x = 9.819$ (Requires quadratic equation.)

Question 4.14: $x = 5000$.

Question 4.15: $\frac{dD}{dp} = -2.4e^{-0.03p}$ and the point for revenue maximization is $p = \frac{1}{0.03} = 33.33\overline{3}$

Question 4.16: $70

Question 4.17: $254.81 (Rounded to the nearest hundredth.)

Question 4.18: $\frac{100}{\sqrt{1200}} = 2.88$.

A.5 ANSWERS TO CHAPTER 5

Question 5.1:

(a) $x = 0$ local maximum.

(b) $x = 2$ local minimum.

(c) $x = -7$ local maximum.

(d) $x = -\frac{7}{2}$ local minimum.

(e) $x = -2$ max, $x = 2$ min

(f) $x = 0$ max, $x = \frac{2}{3}$ min

(g) $x = 0$ min, $x = 2$ max

(h) $x = 0$ max, $x = 2$ min

(i) $x = 0$ max, $x = -2, 2$ mins

(j) $x = \frac{5}{4}$ local maximum.

(k) $x = \frac{1}{4}$ local maximum.

(l) $x = \frac{3}{2}$ local minimum.

Question 5.2:

(a) $(0, 3)$

(b) $(3, 33)$

(c) $(-1, -7), (0, 0), (1, 7)$

(d) $(-2, -80), (2, -80)$

(e) $(\frac{3}{2}, \frac{9}{2})$

(f) $(0, 0)$

(g) $(0, 7)$

(h) $(3, 3)$

(i) $(-1, -21)$

(j) $(0, 1)$

(k) $(4, -105)$

(l) $(0, 12), (2, 4)$

Question 5.3:

(a) D_1: $x = 2$ is absolute minima $x = 4$ is absolute maximum.

  D_2: $x = 2$ absolute minima and $x = -2$ is absolute maxima.

(b) D_1: $x = -1$ is absolute minima $x = 2$ absolute maxima.

  D_2: $x = \frac{2}{3}$ is absolute minima and $x = 2$ is both absolute maxima

(c) D_1: $x = -1$ is absolute minima and $x = 0$ is absolute maxima.

  D_2: $x = 1$ is absolute minima and $x = 2$ is absolute maxima

(d) D_1: $x = 0$ absolute minima and $x = 2$ is absolute maxima.

  D_2: $x = -2$ absolute minima and $x = 1$ absolute maxima.

Question 5.4: $x = 1$ is a local maxima and $x = 4$ is a local minima.

Question 5.5: 36, 35.91

Question 5.6: 500, 500

Question 5.7: 28.48, 28.5

Question 5.8: 5, 0, -5

Question 5.9: 250, 245

Question 5.10: $x = 450$, $P(450) = 6125$, $p = 37.50$

Question 5.11: $x = 525$, $P(525) = 12562.50$, $p = 147.50$

Question 5.12: $x = 187.5 \approx 188$, $P(188) = 10062.40$, $p = 224.80$

Question 5.13: 4000

Question 5.14: 1200

Question 5.15:

(a) -50 (b) -200 (c) -250

Question 5.16: 3

Question 5.17: 2

Question 5.18:

(a) 4375   (b) 87.5   (c) 125

Question 5.19:

(a) $35x - 0.05x^2$ (c) 10   (e) 15

(b) 1000   (d) 5000

Question 5.20:

(a) $MC(x) = 9x^2 - 12x - 8$

(b) $AC(x) = 3x^2 - 6x - 8$

(c) $x = 1.821$ with a local minimum of approximately -13.349.
  $x = -0.488$ with a local maximum of approximately 2.126.

Question 5.21:

(a) $p = 4$ and $x = 4$

(b) $TR(x) = 8x - x^2$ and $MR(x) = 8 - 2x$

Question 5.22:

(a) $q = 25.58$ (To the nearest hundredth.)

(b) The maximum profit is $P(25.58) = 225.32$ (To the nearest hundredth.)

(c) The monopolist price is $p = 17.21$.

Question 5.23: The point of diminishing returns comes into play when $x = 10$.

Question 5.24:

1. The point of diminishing returns occurs when $x = 100$ and sales are $N(100) = 121,800$.

2. Since the marginal sales is diminishing at $x = 100$, spending more on advertising may not result in a proportional increase in sales.

A.6   ANSWERS TO CHAPTER 6

Question 6.1:

(a) $\frac{x^4}{4} + \frac{5x^3}{3} + 3x^2 + c$

(d) $\frac{x^7}{7} - \frac{x^6}{6} + c$

(g) $a^2 \ln|x| + c$

(b) $-\frac{x^3}{3} - \frac{x^4}{4} - 3x + c$

(e) $\int (x-2)\,dx = \frac{x^2}{2} - 2x + c$

(h) $\ln(1+x^2) + c$

(c) $\frac{x^5}{5} - \frac{x^4}{4} + 2x + c$

(f) $\frac{5^x}{\ln 5} + \frac{x^3}{3} - e^x + c$

(i) $\frac{1}{3}\ln|3x+1| + c$

Question 6.2:

(a) $\frac{1}{4} + \frac{5}{3} + 3 = \frac{59}{12}$

(b) $-\frac{20}{3}$

(c) $\frac{104}{5}$

(d) $\frac{256}{7}$

(e) $\frac{1}{2}$

(f) $\frac{4}{\ln 5} + \frac{1}{3} - (e-1)$.

(g) $\int_{-2}^{1} \frac{a^2}{x}\,dx$ is undefined due to division by zero at $x = 0$.

(h) $\ln(10)$

(i) $\frac{\ln(10)}{3}$

Question 6.3: The area enclosed by $f(x) = 4x - x^2$, the x-axis, $x = 0$, and $x = 2$ is $\frac{16}{3}$ or approximately 5.33 square units.

Question 6.4: The area enclosed by $f(x) = x^3 - x^2 - 5x$, the x-axis, $x = -1$, and $x = 1$ is $\frac{9}{2}$ or exactly 4.5 square units.

Question 6.5: The area enclosed by $f(x) = x^3$ and $g(x) = x$ between $x = -1$ and $x = 1$ is $\frac{1}{2}$ or exactly 0.5 square units.

Question 6.6:

1. The revenue function is $R(x) = 40x - 0.01x^2 + c$. Since $R(x) = 0$ when $x = 0$, $c = 0$, so $R(x) = 40x - 0.01x^2$.

2. The revenue from the sale of 1500 pairs of shoes is $R(1500) = \$37,500$.

Question 6.7:

1. The revenue function is $R(q) = 400q - 0.2q^2 + c$. Given $R(q) = 0$ when $q = 0$, $c = 0$, thus $R(q) = 400q - 0.2q^2$.

2. The revenue at a production level of 500 units is $R(500) = 400 * 500 - 0.2 * 500^2 = 200,000 - 500,000 = 150,000$. At 500 units, the revenue is positive and substantial at 150,000.

Question 6.8:

1. The cost function is $C(q) = 0.2q^3 + q^2 + c$. With fixed costs of \$2000, $c = 2000$, so $C(q) = 0.2q^3 + q^2 + 2000$.

2. The cost of producing 25 units is $C(25) = \$5750$.

Question 6.9:

1. The cost function $C(x)$ is $\frac{4}{3}x^3 - \frac{25}{2}x^2 + 50x + c$. With fixed costs of \$30,000, $c = 30,000$, thus $C(x) = \frac{4}{3}x^3 - \frac{25}{2}x^2 + 50x + 30,000$.

2. The cost of producing 4000 bottles (or 4 thousand bottles, since x is in thousands) is $C(4) = \$30,085.33$.

Question 6.10: The total cost function is $C(y) = 75000\ln(y) + \frac{y^2}{2} + c$, assuming $y \geq 0$ since logarithms of non-positive numbers are undefined.

Question 6.11: The total cost function is $C(x) = 6x^2 + 25000\ \ln(x) + c$, where c is the integration constant.

Question 6.12: The total cost function is $C(x) = 2x^2 + 150 \ln(x) + c$, where c is the integration constant.

Question 6.13: The total revenue is $R(x) = -1.5 \times 10^6 e^{-0.0002x} + c$, where c is the integration constant.

Question 6.14: The total revenue from $t = 0$ to $t = 20$ days is given by the integral of $R(t) = P(t) \times Q(t) = (500 - 10t)(100t - 5t^2) = 50000t - 3500t^2 + 50t^3$. To find the total revenue during this period, evaluate this expression from $t = 0$ to $t = 20$, Total Revenue $= \int_0^{20} (50000t - 3500t^2 + 50t^3) \, dt = \left[\frac{50000}{2}t^2 - \frac{3500}{3}t^3 + \frac{50}{4}t^4 \right]_0^{20} = \$2,666,666.67$.

Question 6.15: The profit function $P(x)$ is found by integrating the marginal profit, which is $P'(x) = R'(x) - C'(x)$. Given $C'(x) = 80x$ and $R'(x) = -0.03x^2 + 120x$, the profit function is $P'(x) = (-0.03x^2 + 120x) - 80x = -0.03x^2 + 40x$. Integrating $P'(x)$ to find $P(x)$: $P(x) = \int P'(x) \, dx = \int (-0.03x^2 + 40x) \, dx = -0.01x^3 + 20x^2 + c$

Question 6.16: The consumer surplus when $p = 8 - x$ and $x_0 = 4$ is 8.

Question 6.17: First, we find the Equilibrium by setting the Demand and Supply Equations equal we find $x \approx 2.37$ (since x must be non-negative). Therefore $p \approx 2 + 2.37 = 4.37$. The producer surplus is the area above the supply curve and below the price line from 0 to $x_0 = 2.37$: Producer Surplus $\approx 2.37 \times 4.37 - \int_0^{2.37} (2 + x) \, dx \approx 10.36 - 7.55 = 2.81$

Question 6.18: The consumer surplus given the demand function $p = 12 - 2x$ and $x_0 = 4$ is 8.

Question 6.19: The consumer surplus, given the demand function $y = 12 - 2x$ and the market equilibrium quantity $x_0 = 3$, is calculated as follows: Consumer Surplus $= \int_0^3 (12 - 2x) \, dx - 3 \times (12 - 2 \times 3) = 9$ Therefore, the consumer surplus is 9.

Question 6.20:

1. The equilibrium price p_0 is $\frac{14}{3}$ or approximately 4.67 dollars, and the equilibrium quantity q_0 is $\frac{32}{3}$ or approximately 10.67 units.
2. The maximum price consumers are willing to pay is $m_0 = 10$, thus Consumer Surplus $= \int_{4.67}^{10} (20 - 2p) \, dp = 28.41$. Notice, we could have also computed this as CS $= \int_0^{10.67} \frac{20-x}{2} - 4.67 \, dx$.
3. The minimum price producers are willing to produce products for is $c_0 = 2$, thus Producer Surplus $= \int_2^{4.67} (4p - 8) \, dp = 14.26$. Notice, we could have also computed this as PS $= \int_0^{10.68} 4.67 - \frac{x+8}{4} \, dx$

A.7 ANSWERS TO CHAPTER 7

Question 7.1: $\int_0^2 e^{3x}\, dx = \left[\frac{1}{3}e^{3x}\right]_0^2 = \frac{1}{3}(e^6 - 1)$

Question 7.2: $u = 2x + 1 \implies du = 2dx \implies dx = \frac{du}{2}$

$$\int \frac{1}{2x+1}\, dx = \int \frac{1}{u}\cdot \frac{du}{2} = \frac{1}{2}\log|u| + c = \frac{1}{2}\log|2x+1| + c$$

Question 7.3: $\int \frac{x^2-1}{x^3-x}\, dx = \int \frac{x^2-1}{x(x^2-1)}\, dx = \int \frac{1}{x}\, dx = \log|x| + c$

Question 7.4:

(a) Let $u = x^2 + 1 \Rightarrow du = 2x\, dx \Rightarrow dx = \frac{du}{2x}$,

$$\int \frac{2x}{x^2+1}\, dx = \int \frac{1}{u}\, du = \ln|u| + c = \ln|x^2 + 1| + c$$

(b) $x^2 - 4x + 3 = (x-1)(x-3) \Rightarrow \frac{x-1}{x^2-4x+3} = \frac{A}{x-1} + \frac{B}{x-3}$. Solve for A and B to find: $A = 1, B = -1$,

$$\int \frac{1}{x-1} + \frac{-1}{x-3}\, dx = \ln|x-1| - \ln|x-3| + c$$

(c) First we decompose into: $\frac{A}{x+1} + \frac{Bx+c}{x^2+1}$, then we solve to find $A = 2, B = 2, C = 0$. Thus, the integral simplifies to:

$$\int \frac{2}{x+1} + \frac{2x}{x^2+1} = \ln|x+1| + \ln|x^2+1| + c$$

(d) Use integration by parts with $u = x^2$ and $dv = e^x\, dx$ to find: $x^2 e^x - 2\int xe^x\, dx$. Then one more integration by parts with: $u = x$ and $dv = e^x\, dx$ to find: $x^2 e^x - 2(xe^x - e^x) + c$

(e) Integration by parts with: $u = \ln(x)$ and $dv = x\, dx$ to find: $\int x\ln(x)\, dx = \frac{x^2}{2}\ln(x) - \frac{x^2}{4} + c$

(f) Integration by substitution with: $u = x^2 + 1 \Rightarrow du = 2x\, dx \Rightarrow dx = \frac{du}{2x}$,

$$\int xe^{x^2}\, dx = \frac{1}{2}\int e^u\, du = \frac{1}{2}e^{x^2} + c$$

(g) Integration by substitution: $u = 5 - x^2 \Rightarrow du = -2x\, dx \Rightarrow dx = \frac{du}{-2x}$,

$$\int x\sqrt{5-x^2}\, dx = \frac{-1}{2}\int \sqrt{u}\, du = \frac{-1}{3}(5-x^2)^{\frac{3}{2}} + c$$

(h) Integration by substitution: $u = 3 - x^2 \Rightarrow du = -2x\, dx \Rightarrow dx = \frac{du}{-2x}$,

$$\int \frac{x}{3-x^2}\, dx = \frac{-1}{2}\int \frac{1}{u}\, du = \frac{-1}{2}\ln|3-x^2| + c$$

(i) By substitution with: $u = 2 - x, du = -dx$, so $x = 2 - u$ and $dx = -du$,

$$\int \frac{x}{2-x} dx = -\int \frac{2-u}{u} du = -\int (\frac{2}{u} - 1) du = -(2\ln|2-x| - (2-x)) + c$$

(j) We integration by parts with $u = x$ and $dv = e^{-3x}$,

$$\int xe^{-3x} dx = \frac{-1}{3} xe^{-3x} - \frac{1}{9} e^{-3x} + c$$

(k) Using integration by parts, let: $u = \ln x$ and $dv = (3x+1)dx$,

$$\int (3x+1) \ln x \, dx = (\ln x) \left(\frac{3}{2} x^2 + x \right) - \left(\frac{3}{4} x^2 + x + c \right)$$

(l) Complete the square: $x^2 + 12x + 36 = (x+6)^2$ then use substitution,

$$\int \frac{12000}{x^2 + 20x + 100} dx = -\frac{12000}{x+6} + c$$

(m) Complete the square: $x^2 + 12x + 36 = (x+10)^2$ then use substitution,

$$\int \frac{20000}{x^2 + 20x + 100} dx = -\frac{20000}{x+10} + c$$

Question 7.5:

(a) Using substitution,

$$\int_0^1 \frac{3x}{x^2+4} dx = 3 \left[\frac{1}{2} \ln(x^2 + 4) \right]_0^1 = \frac{3}{2} \left(\ln(5) - \ln(4) \right) = \frac{3}{2} \ln \left(\frac{5}{4} \right)$$

(b) Factor the denominator,

$$\int_1^3 \frac{x-2}{(x-1)(x-2)} dx = \int_1^3 \frac{1}{x-1} dx = \left[\ln(x-1) \right]_1^3 = \ln(2)$$

(c) First we decompose into: $\frac{A}{x+1} + \frac{Bx+c}{x^2+4}$, then we solve to find $A = 1, B = -2, C = 0$. Thus, the integral simplifies to:

$$\int_0^1 \frac{1}{x+1} - \frac{2x}{x^2+4} dx = \left[\ln|x+1| - \ln|x^2+4| \right]_0^1 = \ln(2) - \ln(5) + \ln(4) = \ln \left(\frac{8}{5} \right)$$

(d) Use integration by parts,

$$\int_0^2 x^3 e^x \, dx = \left[x^3 e^x - 3x^2 e^x + 6xe^x - 6e^x \right]_0^2 = 2e^2 + 6$$

(e) Use integration by parts: $u = \ln x, dv = x^2\, dx \Rightarrow du = \frac{1}{x}\, dx, v = \frac{x^3}{3}$

$$\int_1^2 \ln(x)\cdot x^2\, dx = \left[\ln(x)\frac{x^3}{3}\right]_1^2 - \frac{1}{3}\int_1^2 x^2\, dx = \ln(2)\frac{8}{3} - \frac{1}{3}\left[\frac{x^3}{3}\right]_1^2 = \frac{8}{3}\ln(2) - \frac{7}{9}$$

(f) By substitution: $u = -x^2$. Then, $du = -2x\, dx$, $\frac{-du}{2} = x\, dx$. We obtain:

$$\frac{-1}{2}\int_0^1 e^u\, du = \frac{-1}{2}[e^u]_0^1 = \frac{-1}{2}\left(e^{-1} - 1\right) = \frac{-1}{2}\left(\frac{1}{e} - 1\right)$$

(g) Integration by substitution: $u = 5 - x^2 \Rightarrow du = -2x\, dx \Rightarrow dx = \frac{du}{-2x}$.

$$\int_0^{\sqrt{5}} x\sqrt{5-x^2}\, dx = \frac{-1}{2}\int \sqrt{u}\, du = \left[\frac{-1}{3}(5-x^2)^{\frac{3}{2}}\right]_0^{\sqrt{5}} = \frac{5\sqrt{5}}{3}$$

(h) By substitution $u = 3 - x^2$, we get:

$$\int_{-1}^1 \frac{x}{(3-x^2)^2}\, dx = \frac{-1}{2}\int \frac{1}{u^2}\, du = \frac{1}{2}\left[\frac{1}{u}\right] = \frac{1}{2}\left[\frac{1}{3-x^2}\right]_{-1}^1 = 0$$

(i) Let $u = 2 - x, du = -dx$. When $x = 0, u = 2$, and when $x = 1, u = 1$.

$$\int_0^1 \frac{x}{(2-x)^2}\, dx = -\int_2^1 \frac{2-u}{u^2}\, du = \int_2^1 \frac{2-u}{u^2}(-du) = \int_1^2 \frac{2-u}{u^2}\, du$$

$$= \int_1^2 \frac{2}{u^2}\, du - \int_1^2 \frac{1}{u}\, du = \left[-\frac{2}{u} - \ln|u|\right]_1^2 = 1 - \ln(2)$$

(j) Using integration by parts: $u = x$, $dv = e^{-2x}dx$, so $du = dx$, $v = -\frac{1}{2}e^{-2x}$.

$$\int_0^\infty -\frac{1}{2}xe^{-2x}\, dx = \left[-\frac{1}{2}xe^{-2x} + \frac{1}{4}e^{-2x}\right]_0^\infty$$

$$= \lim_{b\to\infty}\left[-\frac{1}{2}be^{-2b} + \frac{1}{4}e^{-2b}\right] - \left[-\frac{1}{2}\cdot 0\cdot e^0 + \frac{1}{4}e^0\right] = \frac{1}{4}$$

(k) Using integration by parts, let: $u = \ln x$ and $dv = (3x+1)dx$

$$\int_1^e (3x+1)\ln x\, dx = \left[\ln x\left(\frac{3}{2}x^2 + x\right) - \left(\frac{3}{4}x^2 + x\right)\right]_1^e = \frac{3e^2}{4} + \frac{7}{4}$$

Question 7.6: The total value of the vehicle after t years is given by $V(t) = 50{,}000 - \int_0^t 5000xe^{-0.2x}\, dx = 125000e^{-0.2t} + 25000te^{-0.2t} - 75000$.

Question 7.7: To determine the remaining value of this equipment after t years, the integration of this function is performed to obtain $V(t) = 200{,}000 - \int_0^t 20000xe^{-0.05x}\, dx = 200000 - (-400000te^{-0.05t} - 8000000e^{-0.05t} + 8000000)$ $= 400000te^{-0.05t} + 8000000e^{-0.05t} - 7800000$

Question 7.8:

(a) $F(x) = \int_0^x \frac{1}{(t+200)^2} \, dt = \frac{1}{200} - \frac{1}{x+200}$

(b) The probability of earning between 500 and 1500 is: $\int_{500}^{1000} \frac{1}{(t+200)^2} \, dt =$
$F(1500) - F(500) = \frac{1}{700} - \frac{1}{1700}$

(c) The probability of earning more than 2000 is $\int_{2000}^{\infty} \frac{1}{(t+200)^2} \, dt = 1 -$
$F(2000) = 1 - \left(\frac{1}{200} - \frac{1}{2200} \right) = \frac{199}{200} + \frac{1}{2200}$

Question 7.9:

(a) $F(x) = \int_0^x \frac{1}{200} e^{-\frac{t}{200}} \, dt = \left[-e^{-\frac{t}{200}} \right]_0^x = 1 - e^{-\frac{x}{200}}$

(b) $P(500 \leq X \leq 1500) = F(1500) - F(500) = \left(1 - e^{-\frac{1500}{200}} \right) -$
$\left(1 - e^{-\frac{500}{200}} \right) = e^{-2.5} - e^{-7.5}$

(c) $P(X > 2000) = 1 - F(2000) = 1 - \left(1 - e^{-\frac{2000}{200}} \right) = e^{-10}$

(d) $E(X) = \int_0^{\infty} x f(x) \, dx$. Using integration by parts with $u = x$ and $dv = \frac{1}{200} e^{-\frac{x}{200}} \, dx$, gives us:

$$E(X) = \left[x \left(-200 e^{-\frac{x}{200}} \right) \right]_0^{\infty} + \int_0^{\infty} 200 e^{-\frac{x}{200}} \, dx = 0 + 200 \cdot 200 = 40000$$

A.8   ANSWERS TO CHAPTER 8

Question 8.1:

(a) 7   (c) 19   (e) -2 (g) 2.49

(b) 768   (d) 28   (f) 27.09   (h) 3

Question 8.2:

(a) $f_x(x, y) = 10x - 15y$
$f_y(x, y) = 6y^2 - 15x$

(b) $f_x(x, y) = 18y^2 x + 2$
$f_y(x, y) = 18x^2 y - 10y$

(c) $f_x(x, y) = 18y^2 x + 6$
$f_y(x, y) = 18x^2 y - 16y$

(d) $f_x(x, y) = -12y^2 x + 10x$
$f_y(x, y) = -12x^2 y + 7$

(e) $f_x(x, y) = 8y^2 x - 6x$
$f_y(x, y) = 8x^2 y - 9$

(f) $f_x(x, y) = 24x^2 y + 2$
$f_y(x, y) = 8x^2 - 14y$

(g) $f_x(x, y) = 5y^3 + 12x$
$f_y(x, y) = 15y^2 x + 3$

(h) $f_x(x, y) = 16x^3 - 30y^3 x^2$
$f_y(x, y) = 28y^3 - 30x^3 y^2$

Question 8.3:

(a) $f_x(x, y) = 24x(3x^2 + y^4)^3$
$f_y(x, y) = 16y^3(3x^2 + y^4)^3$

(b) $f_x(x, y) = -105x^2(-5x^3 - 4y^2)^6$
$f_y(x, y) = -56y(-5x^3 - 4y^2)^6$

(c) $f_x(x, y) = 3(x^2 y + y^2 x)^2 (2yx + y^2)$
$f_y(x, y) = 3(x^2 y + y^2 x)^2 (x^2 + 2xy)$

(d) $f_x(x, y) = 5(3x^2 + 5x^3y^3 - 2y^2)^4(6x + 15y^3x^2)$
$f_y(x, y) = 5(3x^2 + 5x^3y^3 - 2y^2)^4(15x^3y^2 - 4y)$

(e) $f_x(x, y) = -\frac{24x^2}{(4x^3+3y^4)^3}$
$f_y(x, y) = -\frac{24y^3}{(4x^3+3y^4)^3}$

(f) $f_x(x, y) = \frac{8(3x-y^2)}{x^5y^4(-3x+2y^2)^5}$
$f_y(x, y) = \frac{12(x-2y^2)}{x^4y^5(-3x+2y^2)^5}$

(g) $f_x(x, y) = -\frac{25x^4}{2(5x^5-2y^7)^3}$
$f_y(x, y) = \frac{7y^6}{(-2y^7+5x^5)^3}$

(h) $f_x(x, y) = -\frac{6x}{(x^2-y^2)^4}$
$f_y(x, y) = \frac{6x}{(x^2-y^2)^4}$

Question 8.4:

(a) $f_x(x, y) = ye^{xy}$
$f_y(x, y) = xe^{xy}$

(b) $f_x(x, y) = e^{x-y}$
$f_y(x, y) = -e^{x-y}$

(c) $f_x(x, y) = 10xy^3e^{x^2y^3}$
$f_y(x, y) = 15x^2y^2e^{x^2y^3}$

(d) $f_x(x, y) = 9x^2e^{x^3+y^2}$
$f_y(x, y) = 6ye^{x^3+y^2}$

(e) $f_x(x, y) = \frac{1}{x}$
$f_y(x, y) = \frac{1}{y}$

(f) $f_x(x, y) = \frac{6x}{2y^3+3x^2}$
$f_y(x, y) = \frac{6y^2}{2y^3+3x^2}$

(g) $f_x(x, y) = \frac{7}{x-y}$
$f_y(x, y) = -\frac{7}{x-y}$

(h) $f_x(x, y) = \frac{30}{x}$
$f_y(x, y) = \frac{30}{y}$

Question 8.5:

(a) $f_x(x, y) = 2x - 2y$
$f_y(x, y) = -2x + 2y$

(b) $f_x(x, y) = -4y^3 + 18x^2 + 6xy$
$f_y(x, y) = -8y^3 + 3x^2 - 12y^2x$

(c) $f_x(x, y) = 60x^2 + 50xy - 100x$
$f_y(x, y) = 25x^2$

(d) $f_x(x, y) = 60x^2 + 60x^4y^2 - 24x^3y^2 - 20x$
$f_y(x, y) = 24x^5y - 12x^4y$

(e) $f_x(x, y) = -\frac{y}{x^2}$
$f_y(x, y) = \frac{1}{x}$

(f) $f_x(x, y) = 0$
$f_y(x, y) = 0$

(g) $f_x(x, y) = \frac{2x(5x+4y)}{(5x+2y)^2}$
$f_y(x, y) = -\frac{4x^2}{(5x+2y)^2}$

(h) $f_x(x, y) = -\frac{x-3y}{2x^3y^3}$
$f_y(x, y) = \frac{3(y-x)}{2y^4x^2}$

Question 8.6:

(a) $\frac{\partial^2 f}{\partial x^2} = 30y^5x$, $\frac{\partial^2 f}{\partial y\partial x} = 75y^4x^2$, $\frac{\partial^2 f}{\partial x\partial y} = 75y^4x^2$, $\frac{\partial^2 f}{\partial y^2} = 100x^3y^3$

(b) $\frac{\partial^2 f}{\partial x^2} = 126x^5$, $\frac{\partial^2 f}{\partial y\partial x} = 0$, $\frac{\partial^2 f}{\partial x\partial y} = 0$, $\frac{\partial^2 f}{\partial y^2} = -80y^3$

(c) $\frac{\partial^2 f}{\partial x^2} = 120x^4 - 120xy$, $\frac{\partial^2 f}{\partial y\partial x} = -60x^2$, $\frac{\partial^2 f}{\partial x\partial y} = -60x^2$, $\frac{\partial^2 f}{\partial y^2} = 50$

(d) $\frac{\partial^2 f}{\partial x^2} = 6(x^2+y^2)(5x^2+y^2)$, $\frac{\partial^2 f}{\partial y\partial x} = 24xy(x^2+y^2)$, $\frac{\partial^2 f}{\partial x\partial y} = 24xy(x^2+y^2)$,
$\frac{\partial^2 f}{\partial y^2} = 6(x^2 + y^2)(x^2 + 5y^2)$

(e) $\frac{\partial^2 f}{\partial x^2} = 6(9x^2 + y^3)(3x^2 - y^3)^{-3}$, $\frac{\partial^2 f}{\partial y \partial x} = 36xy^2(3x^2 - y^3)^{-3}$, $\frac{\partial^2 f}{\partial x \partial y} =$
$36xy^2(3x^2 - y^3)^{-3}$, $\frac{\partial^2 f}{\partial y^2} = 6y(3x^2 + 2y^3)(3x^2 - y^3)^{-3}$

(f) $\frac{\partial^2 f}{\partial x^2} = 2(x + 3y^2)^{-3}$, $\frac{\partial^2 f}{\partial y \partial x} = 12y(x + 3y^2)^{-3}$, $\frac{\partial^2 f}{\partial x \partial y} = 12y(x + 3y^2)^{-3}$,
$\frac{\partial^2 f}{\partial y^2} = 6(-x + 9y^2)(x + 3y^2)^{-3}$

(g) $\frac{\partial^2 f}{\partial x^2} = e^{x^2 y^2} \cdot (2y^2 + 4x^2 y^4)$, $\frac{\partial^2 f}{\partial y \partial x} = e^{x^2 y^2} \cdot 4xy(1 + x^2 y^2)$, $\frac{\partial^2 f}{\partial x \partial y} = e^{x^2 y^2} \cdot$
$4xy(1 + x^2 y^2)$, $\frac{\partial^2 f}{\partial y^2} = e^{x^2 y^2} \cdot (2x^2 + 4x^4 y^2)$

(h) $\frac{\partial^2 f}{\partial x^2} = e^{x^2 + y^2} \cdot (2 + 4x^2)$, $\frac{\partial^2 f}{\partial y \partial x} = e^{x^2 + y^2} \cdot 4xy$, $\frac{\partial^2 f}{\partial x \partial y} = e^{x^2 + y^2} \cdot 4xy$,
$\frac{\partial^2 f}{\partial y^2} = e^{x^2 + y^2} \cdot (2 + 4y^2)$

(i) $\frac{\partial^2 f}{\partial x^2} = -\frac{3}{x^2}$, $\frac{\partial^2 f}{\partial y \partial x} = 0$, $\frac{\partial^2 f}{\partial x \partial y} = 0$, $\frac{\partial^2 f}{\partial y^2} = -\frac{3}{y^2}$

(j) $\frac{\partial^2 f}{\partial x^2} = \frac{12x(y^3 - x^3)}{(2x^3 + y^3)^2}$, $\frac{\partial^2 f}{\partial y \partial x} = \frac{-18x^2 y^2}{(2x^3 + y^3)^2}$, $\frac{\partial^2 f}{\partial x \partial y} = \frac{-18x^2 y^2}{(2x^3 + y^3)^2}$, $\frac{\partial^2 f}{\partial y^2} = \frac{3y(4x^3 - y^3)}{(2x^3 + y^3)^2}$

Question 8.7:

1. $dz = \frac{\partial z}{\partial x} dx + \frac{\partial z}{\partial y} dy = 12x^2(2x^3 - y^2)dx - 4y(2x^3 - y^2)dy$

2. If x decreases by 3% and y increases by 2%, then $\Delta x = -0.03x$ and $\Delta y = 0.02y$, and the approximate change in z is:

$$\Delta z \approx 12x^2(2x^3 - y^2)\Delta x - 4y(2x^3 - y^2)\Delta y$$
$$= 12x^2(2x^3 - y^2)(-0.03x) - 4y(2x^3 - y^2)(0.02y)$$
$$= -(0.36x^3 + 0.08y^2)(2x^3 - y^2)$$

Question 8.8:

1. $dz = \frac{\partial z}{\partial x} dx + \frac{\partial z}{\partial y} dy = 3e^{x^3 + y^2}(3x^2 dx + 2y dy)$

2. If x increases by 2% and y decreases by 4%, then $\Delta x = 0.02x$ and $\Delta y = -0.04y$, and the approximate change in z is: $\Delta z \approx \frac{\partial z}{\partial x} \Delta x + \frac{\partial z}{\partial y} \Delta y$

$$\Delta z \approx 3e^{x^3 + y^2}(3x^2 \cdot 0.02x + 2y \cdot (-0.04y)) = e^{x^3 + y^2}(0.18x^3 - 0.24y^2)$$

Question 8.9: $\frac{\partial z}{\partial x} = y - 6x^2 y^2 + 3$, $\frac{\partial z}{\partial y} = x - 4x^3 y$

Question 8.10: $\frac{\partial z}{\partial x} = x^x (\ln x + 1) - \frac{3}{x^2}$, $\frac{\partial z}{\partial y} = -4y$

Question 8.11: The profit function $P(x, y)$ can be defined as the difference between the revenue $R(x, y)$ and the cost $C(x, y)$:

$$P(x, y) = R(x, y) - C(x, y)$$
$$= (50x + 70y + 0.07xy - 0.04x^2 - 0.04y^2) - (7x + 9y + 18000)$$
$$= 43x + 61y + 0.07xy - 0.04x^2 - 0.04y^2 - 18000$$

The partial derivatives of the profit function with respect to x and y are:

$$P_x(x, y) = 43 + 0.07y - 0.08x$$
$$P_y(x, y) = 61 + 0.07x - 0.08y$$

Evaluating these at $x = 800$ and $y = 1000$:

$$P_x(800, 1000) = 43 + 0.07 \cdot 1000 - 0.08 \cdot 800 = 49$$
$$P_y(800, 1000) = 61 + 0.07 \cdot 800 - 0.08 \cdot 1000 = 37$$

Question 8.12: The revenue function $R(x, y)$ is:

$$R(x, y) = (250 - 2x + 2y)x + (300 + 3x - y)y$$

$$R(x, y) = -2x^2 - y^2 + 5xy + 250x + 300y$$

The profit function $P(x, y)$, derived by subtracting the cost from revenue, is:

$$P(x, y) = R(x, y) - C(x, y)$$

$$P(x, y) = -2x^2 - y^2 + 5xy + 210x + 100y - 400$$

Calculating partial derivatives for $P_x(x, y)$ and $P_y(x, y)$:

$$P_x(x, y) = -4x + 5y + 210$$

$$P_y(x, y) = -2y + 5x + 100$$

Evaluating these derivatives at $x = 300$ and $y = 200$:

$$P_x(300, 200) = 10$$

$$P_y(300, 200) = 1200$$

Indicates that increasing X by one unit increases profit by \$10, while increasing Y by one unit increases profit by \$1200 at these production levels.

A.9   ANSWERS TO CHAPTER 9

Question 9.1:

(a) $3y^2 \frac{dy}{dx} - 6x - 6 = 0$
$\Rightarrow \frac{dy}{dx} = \frac{6x+6}{3y^2} = \frac{2x+2}{y^2}$

(b) $2y^2 \frac{dy}{dx} - 6\frac{dy}{dx} - 6x = 0$
$\Rightarrow \frac{dy}{dx} = \frac{6x}{2y^2-6}$

(c) $2y\frac{dy}{dx} + 2x = 0$
$\Rightarrow \frac{dy}{dx} = -\frac{x}{y}$

(d) $6y + 6x\frac{dy}{dx} - 6x^2 = 0$
$\Rightarrow \frac{dy}{dx} = \frac{6x^2-6y}{6x} = \frac{x^2-y}{x}$

(e) $3x^3 \frac{dy}{dx} + 9x^2 y - 12y\frac{dy}{dx} - 5 = 0$
$\Rightarrow \frac{dy}{dx} = \frac{5-9x^2 y}{3x^3-12y}$

(f) $1 - \frac{1}{2}y^2 - yx\frac{dy}{dx} - 5\frac{dy}{dx} = 12x$
$\Rightarrow \frac{dy}{dx} = -\frac{12x+\frac{1}{2}y^2-1}{yx+5}$

(g) $-3x^2 - 6y\frac{dy}{dx} = 0$
$\Rightarrow \frac{dy}{dx} = -\frac{x^2}{2y}$

(h) $\frac{6x\left(2x-\frac{dy}{dx}\right)-(x^2-y)(6)}{(6x)^2} = 0$
$\Rightarrow \frac{dy}{dx} = \frac{x^2+y}{x}$

(i) $\frac{2}{\sqrt{x+y}}(1 + \frac{dy}{dx}) - 10x = 0$
$\Rightarrow \frac{dy}{dx} = 5x\sqrt{x+y} - 1$

Question 9.2:

(a) $\lim_{x\to-4}\frac{4-x^2}{x+4}$ Does not exist (Not in indeterminate form so L'Hôpital's rule does not apply.)

(b) $\lim_{x\to0}\frac{x^3-3x}{2x} = -\frac{3}{2}$

(c) $\lim_{x\to2}\frac{x^2-5x+6}{x-2} = -1$

(d) $\lim_{x\to-1}\frac{x+1}{x^3+1} = \frac{1}{3}$

(e) $\lim_{x\to-2}\frac{(x+2)^4}{x+2} = 0$

(f) $\lim_{x\to\infty}\frac{2x^3-4x+2}{x^4+2x^2-5x} = 0$

Question 9.3

(a) $\epsilon(10) = \frac{-1}{11}$. Inelastic.
(b) $\epsilon(10) = \frac{-5}{3}$. Elastic.

(c) $\epsilon(10) = \frac{-3}{2}$. Elastic.
(d) $\epsilon(10) = \frac{-5}{7}$. Inelastic.

Question 9.4 Given the price-demand function: $50p + 10x - 600 = 0$.

(a) The demand x as a function of the price p

$$50p + 10x - 600 = 0$$
$$10x = 600 - 50p$$
$$x = 60 - 5p$$

(b) The elasticity of demand when $p = 5$:

$$D(p) = x = 60 - 5p$$
$$D'(p) = -5$$
$$\epsilon(p) = \frac{D'(p)}{D(p)}p = \frac{-5p}{60 - 5p}$$
$$\epsilon(5) = \frac{-25}{35} = -0.714$$

So the elasticity of demand is -0.714, which is inelastic.
The elasticity of demand when $p = 5$ and the price increases by 10%:

$$p_{\text{new}} = 1.1p = 1.1(5) = 5.5$$
$$\epsilon(p) = \frac{-5p_{\text{new}}}{60 - 5p_{\text{new}}}$$
$$\epsilon(5) = \frac{-5(5.5)}{60 - 5(5.5)} = \frac{-27.5}{32.5} = -0.846$$

So the elasticity of demand is -0.846, which is still inelastic. But notice, we are getting closer to the elastic range.

(c) The elasticity of demand when $p = 8$: $\epsilon(25) = -2$, which is elastic.
(d) The elasticity of demand when $p = 8$ and the price increases by 10%:

$$p_{\text{new}} = 1.1p = 1.1(8) = 8.8$$
$$\epsilon(8.8) = \frac{-5(8.8)}{60 - 5(8.8)} = -2.75$$

So the elasticity of demand is -2.75, which is even more elastic.

Question 9.5 Given the price-demand function: $0.4x + p = 20$.

(a) To write demand x as a function of the price p, we can solve the given price-demand equation for x:

$$0.4x + p = 20 \quad \Longrightarrow \quad x = 50 - 2.5p$$

(b) The revenue $R(p)$ is given by the product of price p and demand x:

$$R(p) = px = p(50 - 2.5p) = 50p - 2.5p^2$$

(c) The elasticity of demand $\epsilon(p)$ is given by:

$$\epsilon(p) = \frac{D'(p)}{D(p)}p = \frac{-2.5p}{50 - 2.5p} = \frac{-p}{20 - p}$$

(d) If $-1 < \epsilon(p) < 0$, the demand is inelastic, and if $\epsilon(p) < -1$, the demand is elastic. So for the given price-demand function, the demand is elastic for $p > 10$ and inelastic for $p < 10$.

(e) To find the intervals where revenue is increasing or decreasing, we can take the derivative of the revenue function with respect to price p and look at its sign:

$$\frac{dR}{dp} = 50 - 5p$$

So revenue is increasing for $p < 10$ and decreasing for $p > 10$.

Question 9.6:

(a) $F(\lambda M, \lambda R) = 12(\lambda M)^{1/2}(\lambda R)^{1/2} = 12\lambda M^{1/2}R^{1/2} = \lambda F(M, R)$. This suggests constant returns to scale. Simply put, $\alpha + \beta = 1$ implies constant returns to scale.

(b)

$$MP_M = \frac{\partial F}{\partial M} = \frac{\partial}{\partial M}\left(12M^{1/2}R^{1/2}\right) = 6M^{-1/2}R^{1/2}$$

$$MP_R = \frac{\partial F}{\partial R} = \frac{\partial}{\partial R}\left(12M^{1/2}R^{1/2}\right) = 6M^{1/2}R^{-1/2}$$

(c)

$$MRTS_{M \to R} = \frac{MP_R}{MP_M} = \frac{6M^{1/2}R^{-1/2}}{6M^{-1/2}R^{1/2}} = \frac{M}{R}$$

This ratio indicates the rate at which rennet can substitute milk while maintaining the same level of production. As R increases, more milk can be substituted by rennet, reflecting a reciprocal relationship.

Question 9.7:

(a) $MP_L = \frac{\partial Y}{\partial L} = 0.6AK^{0.4}L^{-0.4}$.

(b) $MP_K = \frac{\partial Y}{\partial K} = 0.4AK^{-0.6}L^{0.6}$.

(c) Increasing labor while holding capital constant will lead to a decrease in the marginal productivity of labor. Conversely, increasing capital leads to an increase in the marginal productivity of labor, as more capital makes each unit of labor more productive. Similarly, increasing labor makes each unit of capital more productive, reflecting an increase in the marginal productivity of capital.

Question 9.8: Given $f(x,y) = x^2 + y^2$ subject to $x + y = 1$, the critical point found using the method of Lagrange multipliers is $x = \frac{1}{2}$, $y = \frac{1}{2}$. To determine whether this point is a maximum or a minimum, we evaluate $f(x,y)$ at this point: $f\left(\frac{1}{2}, \frac{1}{2}\right) = \left(\frac{1}{2}\right)^2 + \left(\frac{1}{2}\right)^2 = \frac{1}{2}$. Since $f(x,y)$ is always positive and increases as x and y move away from the point $\left(\frac{1}{2}, \frac{1}{2}\right)$ within the constraint, this point represents the minimum value of $f(x,y)$ under the given constraint.

Question 9.9: Given $f(x,y) = x^2 y$ subject to $x^2 + y^2 = 1$, the critical points found are:

- $x = 0, y = -1$
- $x = 0, y = 1$

- $x = -\sqrt{\frac{2}{3}}, y = -\frac{1}{\sqrt{3}}$
- $x = -\sqrt{\frac{2}{3}}, y = \frac{1}{\sqrt{3}}$

- $x = \sqrt{\frac{2}{3}}, y = -\frac{1}{\sqrt{3}}$
- $x = \sqrt{\frac{2}{3}}, y = \frac{1}{\sqrt{3}}$

To determine the maximum and minimum values, we evaluate $f(x,y)$ at these points. The maximum value of $f(x,y)$ is obtained at $x = \sqrt{\frac{2}{3}}, y = \frac{1}{\sqrt{3}}$ and $x = -\sqrt{\frac{2}{3}}, y = \frac{1}{\sqrt{3}}$, and the minimum value is obtained at $x = \sqrt{\frac{2}{3}}, y = -\frac{1}{\sqrt{3}}$ and $x = -\sqrt{\frac{2}{3}}, y = -\frac{1}{\sqrt{3}}$.

Question 9.10: The cost-minimizing combinations of x and y that satisfy the production constraint are piecewise defined as follows:

$$\text{For } x: x = \begin{cases} -5 & \text{if } b = 0 \text{ and } a > 0, \\ \frac{-5\sqrt{a^2}}{a} & \text{if } b = 0 \text{ and } a < 0, \\ \frac{-5a}{\sqrt{a^2+b^2}} & \text{if } b \neq 0 \end{cases}$$

$$\text{For } y: y = \begin{cases} -5 & \text{if } b = 0 \text{ and } a = 0, \\ -5\sqrt{\frac{b^2}{a^2+b^2}} & \text{if } b > 0, \\ 5\sqrt{\frac{b^2}{a^2+b^2}} & \text{if } b < 0 \end{cases}$$

Question 9.11: To maximize the profit function subject to the given constraint, consider the corner points of the feasible region defined by the constraint. Since the function is linear and we are dealing with a standard linear programming problem, the maximum occurs at a vertex of the feasible region. The vertices are:

- When $x = 0$, $y = 100$
- When $y = 0$, $x = 100$

Evaluating the profit function at these vertices: $P(0, 100) = 5(0) + 3(100) = 300$ and $P(100, 0) = 5(100) + 3(0) = 500$. Thus, the maximum profit of 500 occurs when $x = 100$ and $y = 0$.

Question 9.12: Given the joint cost function $f(x,y) = x^2 + 2y^2 - xy$ subject to the constraint $x + y = 8$, the company should manufacture 5 devices of type x and 3 devices of type y to minimize its cost.

Question 9.13: The consumer should consume $\frac{27}{4}$ units of good x and $\frac{45}{8}$ units of good y to maximize his utility.

Question 9.14 To find the utility-maximizing combination of x and y, we use the method of Lagrange multipliers. Define the Lagrangian as: $\mathcal{L}(x, y, \lambda) = x^\alpha y^\beta + \lambda(M - p_1 x - p_2 y)$. Take the partial derivatives and set them to zero:

$$\frac{\partial \mathcal{L}}{\partial x} = \alpha x^{\alpha-1} y^\beta - \lambda p_1 = 0$$

$$\frac{\partial \mathcal{L}}{\partial y} = \beta x^\alpha y^{\beta-1} - \lambda p_2 = 0$$

$$\frac{\partial \mathcal{L}}{\partial \lambda} = M - p_1 x - p_2 y = 0$$

Solving these equations, we find:

$$\lambda = \frac{\alpha x^{\alpha-1} y^\beta}{p_1}$$

$$\lambda = \frac{\beta x^\alpha y^{\beta-1}}{p_2}$$

Equating the two expressions for λ: $\frac{\alpha x^{\alpha-1} y^\beta}{p_1} = \frac{\beta x^\alpha y^{\beta-1}}{p_2} \implies \frac{\alpha}{\beta} = \frac{x p_2}{y p_1} \implies x = \frac{\alpha}{\beta} \frac{y p_2}{p_1}$ Substituting x in terms of y into the budget constraint gives: $p_1 \left(\frac{\alpha}{\beta} \frac{y p_2}{p_1} \right) + p_2 y = M \implies \left(\frac{\alpha p_2}{\beta} + p_2 \right) y = M \implies y = \frac{M\beta}{\alpha p_2 + \beta p_2}$ And substituting back to find x: $x = \frac{\alpha}{\beta} \frac{y p_2}{p_1} \implies x = \frac{\alpha M}{\alpha p_1 + \beta p_1}$ Thus, the utility-maximizing quantities of x and y are: $x = \frac{M\alpha}{(\alpha+\beta)p_1}$, $y = \frac{M\beta}{(\alpha+\beta)p_2}$.

Index

For Product Safety Concerns and Information please contact our EU
representative GPSR@taylorandfrancis.com
Taylor & Francis Verlag GmbH, Kaufingerstraße 24, 80331 München, Germany